C. P Kunhardt

Steam Yachts and Launches, Their Machinery and Manegement

A review of the steam engine as applied to yachts; laws governing yachts in American waters: A review of the steam engine as applied to yachts; laws governing yachts in American waters; rules for racing; rules for building; pilot regulations; specifi

C. P Kunhardt

Steam Yachts and Launches, Their Machinery and Manegement

A review of the steam engine as applied to yachts; laws governing yachts in American waters:
A review of the steam engine as applied to yachts; laws governing yachts in American waters;
rules for racing; rules for building; pilot regulations; specifi

ISBN/EAN: 9783337413651

Printed in Europe, USA, Canada, Australia, Japan

Cover: Foto ©berggeist007 / pixelio.de

More available books at www.hansebooks.com

STEAM YACHTS AND LAUNCHES;

THEIR

MACHINERY AND MANAGEMENT.

A REVIEW

OF THE STEAM ENGINE AS APPLIED TO YACHTS; LAWS GOVERNING

YACHTS IN AMERICAN WATERS; RULES FOR RACING; RULES

FOR BUILDING; PILOT REGULATIONS; SPECIFIC

TYPES OF MACHINERY; DESIGN OF HULLS;

ETC., ETC., ETC.

BY

C. P. KUNHARDT.

NEW YORK:
FOREST AND STREAM PUBLISHING CO.
1887.

G

PREFACE.

STEAM YACHTING in America has made rapid strides during the past ten years, the fleet of decked yachts now numbering several hundred, to which may be added nearly a thousand launches and other small craft. The possibilities for future expansion are almost beyond estimation. Not only the extensive coast, but the great fresh water lakes and the vast river systems of the American continent are peculiarly adapted to yachting under steam, whether for hunting and fishing purposes, excursions, the pursuit of mechanical tastes, or for agreeable methods of conveyance.

With the many radical improvements in safety, economy and speed which have characterized the development of machinery in recent years, we may look for constantly augmenting accessions to the steam pleasure fleet, until it shall surpass in number and variety the combined fleets of other nations.

Few new buyers of steam yachts have more than a vague comprehension of the driving power of their vessels, and few have the time or inclination to enter upon a prolonged scientific study of the theory of steam machinery, particularly when the practical results to them do not seem proportional to the efforts put forth. This volume is intended to be sufficiently comprehensive, and elementary at the same time, to suit the yacht owner's object of acquiring a general understanding of the subject as a whole, with specific information and data covering the most recent practice.

C. P. K.

NEW YORK, May 1, 1887.

CONTENTS.

	PAGE.
THEORY OF THE STEAM ENGINE,	9-30

Earliest stages. James Watt and his improvements. Hornblower's compound engine and separate condenser. Saturated steam. Marriotte's law of expansion. Economy of expanding steam. Limit to cutting off in practice. Limit to initial pressure. High range of temperature a necessity for economical working; but low range in each cylinder. Economy of the compound engine in practice compared with theoretical deductions. The direction improvements should take. Steam jacketing. Superheating. Friction of piping. Expansion valves. Record of progress made. Steamer Hassler. Triple expansion engines. Steam yacht Gladiator. Power derivable from the fuel. Jacob Perkins and his experiments. Conclusions.

BOILER EFFICIENCY, - - - - - - - - 30-46

Cylindrical boilers. Requisites of good boilers. Mechanical considerations. Temperature of escaping gases to create draft. Clyde return tubular boilers. Perkins pipe boilers. Evaporative power of coal. Air required for combustion. Forced draft. Blowers. Grate and heating surface. Tubes. Combustion chambers. Locomotive boilers. Vertical tubular boilers. Corrugated fire-box flues. Naphtha and electricity as motors. Mineral oil as fuel. Duty of yacht clubs and yacht owners. Advance in British practice. Americans behind in speed. Highest speed recorded. Quadruple engines and their economy.

BOILER MOUNTINGS, - - - - - - - - 47-68

Hydrokineters to promote circulation. Attachments enumerated. Parts of boiler explained. Safety valve. Smokestacks. Steam gauges. Mercury gauges. Water gauges and cocks. Fire plugs. Low-water alarms. Check valves. Inspirators. Blow-off valves. Salinometers Feed pumps. Expansion joints.

THE ENGINE AND ITS PARTS, - - - - - - 69-95

The slide valve and its operation. Eccentrics. Reversing gear. Indicator and diagram. Horse power. The Pantograph. Condensers. Outboard condensing pipes. Jet condensers. Revolution counters.

THE SCREW, - - - - - - - - - - 96-106

Paddlewheels considered. The feathering wheel. Operation of the screw. Nomenclature of the screw. Pitch. Slip, apparent and real. Experiment the only guide. Resistance at high speeds. Various kinds of screws. Loss of power between cylinder and screw. Formulæ for resistance and power untrustworthy. Comparison the only guide to apportioning driving power. Consumption of fuel at various speeds.

Contents.

LAWS APPLICABLE TO STEAM YACHTS, - - - - 107–128
Status of steam yachts. No license fees for inspection. Special license to yacht engineers, masters and pilots. The Revised Statutes relating to licensing yachts. Boiler plate. Test pressures. Thickness of tubes. Space around boilers. Manholes. Fire plugs. Gauges and safety valves. Seacocks. Lifeboats and equipment. Rafts. Bulkheads. Steam launches. Provisions against fire. Licenses to officers. Engine-room signals. Annual inspection. Lock safety valves. Pilot rules for lake and seaboard. Pilot rules for Western rivers. Lights for steam vessels. Case of the Yosemite (see also Addenda, at end of volume).

EXTRACTS FROM LLOYDS' RULES, - - - - - 129–136
Stern framing. Reverse frames under engines and boilers. Garboard strakes. Bulkheads. Skylights. Coal bunkers. Material for boilers. Stays and rivets. Mountings required. Strength of material. Annual surveys. Characters given in Lloyds' Yacht Register. Scantling for wood and iron yachts. Anchors and chains. Butts and rivet work.

RACING STEAM YACHTS, - - - - - - 137–145
Objects of racing. Time allowance by length and its shortcomings. Emory table of allowance; C. H. Haswell's formula preferable. Length should be included.

MANAGEMENT AND CARE OF MACHINERY, - - - 146–154
Filling the boiler. Lighting fires. Safety valve. Foaming. The feed. Low water. Inspirator fails to feed. Blowing off. Banking fires. Removing sediment. Split tubes. Blisters. General care. Laid up. Getting under way. Attendance while running. Thumping.

PRINCIPAL TYPES OF YACHT MACHINERY.
Perkins high pressure system, - - - - - - 155
Herreshoff system, - - - - - - - 164
Vertical direct-acting engines, - - - - - - 179
Sternwheel boats, - - - - - - - 192
Seagoing launches, - - - - - - - 195
Coasting yachts, - - - - - - - 198
Cruising steamers, - - - - - - - 200
Dimensions of steam yachts, - - - - - 203
Wells balance engine, - - - - - - - 207
Colt disc engine, - - - - - - - 212
Cheap machinery for small craft, - - - - - 218
Shipman kerosene engine, - - - - - - 221
Oscillating engines, - - - - - - - 223
Kane's porcupine boiler, - - - - - - 227
Naphtha launches, - - - - - - - 229

THE DESIGN OF HULLS, - - - - - - 233–236
No precise directions can be laid down. Ballast in steam yachts not usual. Resistance and beam. Proportions determined by experience. Consideration of form of least resistance. Limits to its adoption in steam vessels. Proportions of high-speed torpedo boats. Character of cross-section and waterlines dependent upon individual preference.

ADDENDA, - - - - - - - - 237–239
Lights on steam yachts and useful tables.

ILLUSTRATIONS.

	PAGE.
Steam Yacht Shaugraun	*Frontispiece.*
Inverted Launch Engine	15
Double Direct-acting Engine	17
Compound Inverted Surface Condensing Engine	19
Machinery of Steam Yacht Gladiator	25, 27, 28, 29
Return Tubular Boiler	32
Clyde Boiler	33
Perkins Pipe Boiler	34
Doghouse Boiler	36
Sturtevant Blower	37
Locomotive Boiler	39
Vertical Tubular Boiler	40
Corrugated Flue	41
U. S. Launch Return Tubular Boiler	48, 49
Safety Valve	51
Marine Pop Safety Valve	52
Bourdon Steam Gauge	53, 54
Mercury Gauge	55
Combination Gauges	56
Low Water Alarm	58
Check Valve	59
Hancock Inspirator	60
Salinometer	63
Duplex Feed Pump	65, 66
Plunger Pump	68
Section of Cylinder and Valve	70, 71
Reversing Gear	74
Thompson Indicator	76
Indicator Diagram	78
Pantograph	82, 83
Wheeler's Surface Condenser	87, 88, 89
Worthington Jet Condenser	92
Engine Room Counter	94
Willard's Propeller Wheel	98

Illustrations.

Thorneycroft's Screw..102
Giant Propeller..103
Duncan's Propeller...104
Perkins Condenser and Still..156
Perkins Engine..158, 161, 162
Herreshoff Coil Boiler...169
Herreshoff Engines..174, 175
Stiletto, U. S. N..178
Katrina, Launch..180
Plan of 80 ft. Yacht...181
Open Steam Launch..183
Plans of Triple Screw Launch...184
Mohawk, Steam Launch...185
Plans of Revenue Cutter..186
Plans of Small Trading Steamer..................................187, 188
Willard High Speed Engine.......................................190, 191
Sternwheel Boats..192, 193
Cruising Launch..195
Falcon, Coasting Steam Yacht....................................197, 198
Chemcheck, Cruising Steam Yacht......................................201
Carmen, Seagoing Steam Yacht....................................204, 205
Wells Compound Balanced Engine.......................................209
Colt Disc Engine...213, 214, 215
Shipman Kerosene Engine...219, 221
Shipman Launch...222
Kriebel Oscillating Engine......................................224, 225
Kane's Porcupine Boiler and Engine...................................227
Naphtha Launches...229, 230, 231
Naphtha Launch Engine..232

STEAM YACHT SHAUGRAUN.

STEAM YACHTS AND LAUNCHES.

I.

THEORY OF THE STEAM ENGINE.

IN the first stages of its development, the steam engine was a single-acting affair worked without expansion. That is to say, steam was admitted on one side of the piston only, the whole length of the stroke being followed up by a continuous supply of steam from the boiler. The return stroke was accomplished by shutting off the steam and injecting cold water under the piston, thereby condensing the steam and creating a vacuum in the cylinder. The pressure of the atmosphere, 14.7 lbs. per square inch, acting upon the upper side of the piston, would force it to descend preparatory to the next upward stroke.

Between the years 1775 and 1785, the untiring industry and genius of James Watt had wrought such material improvements upon the original "atmospheric engine" of Newcomen, that the same principles and general arrangements continue to prevail in the standard motor of the present day. Under the name of the Cornish pumping engine, almost the identical process and mechanism of Watt are preserved, and until quite recently this was the most economical of all engines used for drawing water from deep mining shafts. Under Watt's patents relating to working by the expansion of steam in a double-acting engine, in which the steam enters on each side of the piston alternately, the opposite side being in communication with the condenser, the whole range of modern refinements in methods of execution is included. The most important and profitable of all the numerous progressive steps toward perfection originated by Watt was the practical application of working by expansion. Strange to say, the true value of this feature remained almost dormant for a long time, and has only received full recognition of late with the general introduction of high-pressure and compound engines, as the means for utilizing the benefits to be derived from the elastic properties of steam. Although individual leaders in the engineering world never lost sight of this great essential to the economic transmission of heat-energy with steam as the conveyor, the majority rested content with the half-way results achieved through low pressures and a low rate of expansion. Undoubtedly the mechanical difficulties in the way of providing apparatus of sufficient strength to withstand the immense pressures now common had much to do with keeping practice far in the rear of theory and small scale experiments.

At this period there are still some who refuse to concede the manifest advantages of pursuing to the mechanical limit the clear dictates of science, through confounding the primary source of saving in the boiler with the mere mechanism of transmission represented in the engine. Fortunately the actual results obtained from working at high pressures and high rates of expansion have held out so much encouragement that the energy displayed in recent years in the direction clearly pointed out by Watt a century ago is destined to lead up to attainments which would have been considered visionary a decade ago.

The impression that the so-called modern system of "compounding" is a recent invention is as far from the truth as that there is a direct profit in the compound engine. The latter contrivance is almost contemporaneous with the birth of the steam engine itself, and was the natural expression of the value set upon expansion by Watt and the philosophers of his period. In 1781, Jonathan Hornblower brought out his compound or double-cylinder engine, in which the exhaust steam from a small cylinder was passed into a larger chamber for expansion and thence into a separate condenser. Watt claimed priority in the separate condenser and also in this arrangement, on the ground that it was nothing more nor less than working steam expansively, the two cylinders being only a mechanical division of his single cylinder in which steam was cut off at a fraction of the stroke. The claim was a just one, for the compound engine differs from the single expansion cylinder only in arrangement of parts and not at all in point of principle. Many years later Rankine, in his investigations of steam and other prime movers, enunciated the law that, "So far as the theoretical action of the steam on the piston is concerned, it is immaterial whether expansion takes place in one cylinder, or in two or more cylinders."

The true source of increased efficiency in modern practice is to be ascribed to the higher initial or boiler pressures adopted, permitting correspondingly higher rates of expansion ; in other words to profiting from the greater elasticity of steam at greater pressures, as the following considerations will explain.

According to Marriotte's law, the pressure of gases varies inversely as the spaces occupied. If, for example, a gas having 40 lbs. pressure per square inch be expanded to twice its former volume, the pressure per square inch will have been reduced to 20 lbs. If it be allowed to expand to three times the original volume, the pressure will be one-third of 40 or 13.33 lbs., and so on. In practice, there will be a deviation from this theoretical law in the cylinder of a steam engine, owing to the variations in temperature which affect the pressure very materially, unless the steam be superheated, when it acquires the properties of a permanent gas, or unless the temperature of the cylinder can be kept nearly up to that of the steam to prevent condensation of saturated steam.

The term saturated steam applies to steam only sufficiently heated to remain in the shape of vapor, an intermediate state between the fluid and permanently gaseous state. Heating saturated steam converts the small globules of water held in suspension into steam, the whole passing ultimately into the superheated state, if the pressure be not increased. Saturated steam, upon cooling, will be precipitated or condensed into water.

The chief cause for the failure of steam in practice to meet the relations expressed in Marriotte's law is the intermittent chilling of the cylinder on the exhaust side of the piston, owing to communication with the condenser, in which the temperature is necessarily low. When, on the next stroke, live steam is admitted, it finds the metal of the cylinder, its head and piston cooled down to such a degree that partial condensation of the saturated steam takes place with corresponding reduction in pressure as the piston proceeds. After the steam has been cut off, the temperature rapidly falls with the expansion during the remainder of the stroke, bringing the temperature lower than that of the metal. The condensed particles on the surface of the metal are then re-evaporated, absorbing heat from the metal at the end of the stroke which passes off into the condenser as so much additional energy wasted. In spite of these deviations, Marriotte's law is employed in ascertaining the mean pressure on the piston.

Suppose the annexed diagram to represent a cylinder, in which the steam is cut off at half stroke. Divide the cylinder into say ten equal parts. If the initial pressure be taken as 1, the pressure at each of the five divisions up to half stroke will also be represented by 1. At B the steam is cut off and expansion begins. At the sixth division the pressure will be reduced to ⅚ of $1 = .8333$, since the spaces occupied at the fifth and sixth divisions are inversely to one another as 6 to 5. At the seventh division, the pressure will have been reduced to ⁵⁄₇ of $1 = .7143$. At the eighth division to ⅝ of $1 = .6250$, and so on. At the tenth or end of stroke the pressure will be one-half the initial, or 0.5, the space occupied being twice as great as when the steam was cut off. Had the steam port been closed at one-

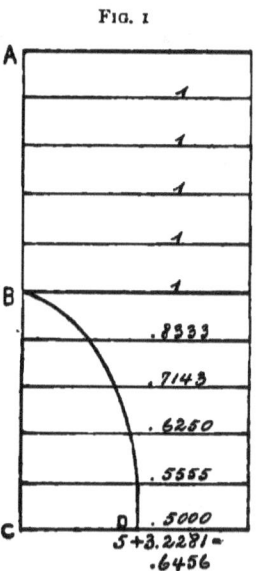
FIG. 1

quarter stroke, the final pressure would have been one-quarter of the initial. The mean of all the ascertained pressures after cutting off is found by adding them and dividing by the number of spaces, as performed in the diagram. The mean of this and the "following steam" for the first half of the stroke is

$$\frac{1+.6456}{2} = .8228,$$

the accuracy of which depends upon the number of divisions introduced.

If the pressures be scaled off on their respective divisions and a curve passed through the extremities, as B D, it will be found to be a hyperbola. From the foregoing we have for the pressure at any point the formula,

$$P' = \frac{P L}{L'}$$

in which P is the initial pressure, L the length of the stroke completed when steam is cut off and L' the length of stroke to the point P'.

Example: The stroke of an engine is 60 inches, and cuts off at 30 inches; initial pressure is 40 lbs.; required the pressure at end of stroke:

$$P' = \frac{30 \times 40}{60} = 20 \text{ lbs.}$$

The effective or "unbalanced" pressure upon the piston will be modified by whatever back pressure may exist on the opposite side of the piston. If the engine were non-condensing, the exhaust would lead into the open air against the atmospheric pressure of 14.7 lbs. to the square inch. The effective pressure at end of stroke in the above example would be $20 - 14.7 = 5.3$ lbs.

If the engine is of the condensing kind, the exhaust would lead into the more or less perfect vacuum of the condenser, and the back pressure would vary accordingly, usually from 2 to 3 lbs., augmented somewhat by the back pressure of the steam itself while escaping on the exhaust side through the pipes connecting with condenser. The mean pressure, calculated from the diagram as .6456, was stated to be only an approximate result. By increasing the number of divisions very greatly, a larger and nearly correct average would be returned in the figure .693, the "hyperbolic logarithm" of 2. Similarly for cutting off at one-quarter stroke, the ratio of pressure for the expansion through the remaining three-quarters of the stroke will be expressed by the hyperbolic logarithm of 4, or 1.386. From a table of such logarithms the ratio of pressure can be obtained for any point of cut-off by simple inspection, doing away with laborious calculation by a large number of divisions in each case. The figures cited by way of example are enough for our purpose.

If the cylinder A C be supposed to measure two units in length, and its cross section one unit in area, then the work performed during the first half stroke, while the piston is forced from A to B by live steam having an initial pressure of 1, will be expressed by $1 \times 1 \times 1 =$ area $\times$ pressure $\times$ distance traveled. The work performed during the latter half of the stroke will be expressed by $1 \times .69 \times 1 = .69$, and the total work throughout the entire stroke by the sum of the two, or $1 + .69 = 1.69$. Had the steam been exhausted at B and not used expansively from B to C, the work performed would have been only 1 instead of 1.69, the quantity of steam used being the same in both cases. Hence, by utilizing the elasticity of the steam in prolonging the stroke after cutting off the supply, the gain in work is represented by 69 per cent. Had the steam port not been closed at all, but had we followed up with live steam for the whole stroke, the work performed would have been $1 \times 1 \times 2$ (area $\times$ pressure $\times$ distance A B C) $= 2$. By cutting off, the work performed was reduced from 2 to 1.69. For drawing fair inferences, however, the work performed should be made equal in both cases. This can be done by increasing the initial pressure when cutting off, so that the mean pressure may be sufficiently great to accomplish the same total work when cutting off at half stroke as with a steady pressure of 1. Call the desired initial pressure X, then $2 : 1.69 :: X : 0.845 X$, (the mean pressure). The mean pressure derived from the new initial pressure sought must of course be the same as the mean pressure when the full stroke is followed up. But this we have assumed to be 1, hence $.845 X = 1$ lb., or $X = 1.18$ lbs., the new initial pressure sought. In cutting off with this initial pressure of 1.18 lbs., only half the cylinder full of steam is used to every full cylinder when following up full stroke with 1 lb. pressure. The difference between 1.18 and 2 therefore repre-

Theory of the Steam Engine. 15

sents the saving in steam by starting with higher initial pressure but cutting off at half stroke. This difference is 0.82 or 41 per cent. This 41 per cent. is the volume of steam saved at the cylinder entrance in performing a certain amount of work or running a certain

FIG. 2.—INVERTED DIRECT-ACTING LAUNCH ENGINE. N. Y. SAFETY STEAM POWER CO.

distance, but it is not identical with the amount of fuel saved, since it requires more coal to raise steam to the higher initial pressure in the first place, with correspondingly increased loss in heat passing out of the chimney to produce the required draft, loss in radiation, friction and condensation in reaching the cylinder. The actual saving in fuel will be found to be about 20 per cent.

The economy of working expansively with increased pressure is evident from the foregoing.

Theoretically there is no limit to the benefits to be derived from expansion. Cutting off at one-seventh would effect double the work done by "following" full stroke, and if we could start with an infinitely great pressure, we could cut off at an infinitely small fraction and effect the maximum saving. As there is always more or less back pressure on the exhaust side, the theoretical limit of cutting off for a given initial pressure would be found at that point which would leave for the terminal pressure at the end of the stroke enough to balance the back pressure with something to spare for overcoming the friction of the engine.

In actual practice this limit is, however, greatly curtailed by a fact already referred to, the lowering of the temperature and consequent loss in power with every increase in expansion. Thus what we would be gaining theoretically by resorting to high rates of expansion may in practice be lost through the accompanying fall in the temperature of the steam. Experience has established that as a rule no economy follows expansion in a single cylinder to a greater number of volumes than one-half the square root of the steam pressure, or algebraically, $\frac{1}{2} \sqrt{P}$, where P is the pressure in pounds per square inch. That is, we should cut off at about $\frac{1}{2}$ for 15 to 20 lbs. pressures, at $\frac{1}{3}$ for say 36 lbs., at $\frac{1}{4}$ for 64 lbs. and $\frac{1}{5}$ for 100 lbs. If we were to expand beyond the fractions determined by experience, the consequence would be a lowering of the mean pressure as explained. This would have to be made good by greater initial or boiler pressure, which implies extra consumption of fuel, and the limits of practical economy would have been overstepped in the pursuit of a theoretical truth.

For a long time steam was treated as a gas of fixed properties, yet every-day routine was demonstrating it to be most unstable and exceedingly sensitive to differences in temperature. It is mainly through the full appreciation of this instability that the recent great advances in the practice have had their birth. The inferences from Marriotte's law were plain enough. Steam should be used at great pressure and at a great rate of expansion. So far as pressure was concerned, it resolved itself into the question of constructing boilers to suit. The mechanical limit to possibilities in construction would be the only restriction to initial pressure. But when it comes to great expansion in the single cylinder, formerly in universal use, practice has been found to fail to meet theory to the extent expected.

The great range in temperature from beginning to end of stroke

FIG. 3.—DOUBLE INVERTED DIRECT-ACTING YACHT ENGINE, WITH INDEPENDENT CUT-OFF. CLUTE & CO., SCHENECTADY, N. Y.

caused such serious losses from condensation and re-evaporation, besides those due to conduction and radiation of the metal, that the mean pressure was sensibly reduced and the gain in economy did not off-set other drawbacks. The revival of Hornblower's method, in other words the introduction of the compound engine, overcomes in a great measure the difficulty experienced in the single cylinder, and supplies us with the mechanical means of profiting from the physical advantage of working with high rate of expansion. We are enabled in the compound arrangement, to achieve the vital desideratum for successful expansion, that is a low *range* of temperature in each cylinder, giving a *high* range for both. Therein lies the superiority of Hornblower's system over the single long chamber required in the simple engine. The process of expansion is cut short in the first cylinder and then finished in the second, so that we have a smaller range of temperature in each than if the expansion had been wholly carried out in one cylinder and consequently less loss of heat energy. The saving thus effected in power represents in turn a saving in fuel, followed by a train of advantages of special application in yacht construction.

With a given amount of fuel, the economical machine will propel the hull to a greater distance, or taking the distance as the basis of comparison, the economical machine can be supplied from smaller coal bunkers. The weight saved can be taken from the yacht's displacement and higher speed brought about by fining away the model, or, what is the same thing, less resistance at the same speed and therefore less expenditure in driving power, thereby making a saving at both ends. The boilers will be smaller and lighter and the same holds good of the engines. High pressures can be introduced with the compound engine with less weight of metal and wear than in the simple cylinder, as the difference between initial and terminal pressure in each cylinder will not be as great and the compound will also be smoother in running. Hence, higher piston speed can be employed, which is equivalent to a further saving by shortening the period during which the metal in contact with the steam can alternately heat and cool. The more rapidly the piston travels, the greater the number of revolutions and the finer the pitch of the

FIG. 4.

A. High pressure cylinder.
B. Low pressure cylinder.
C. Steam from boiler.
D. Valve chests, containing slide valves.
E. F. Connecting rods attached to crossheads P.
G. H. Cranks on shaft.
I. J. Eccentrics, giving motion to valve gear.
K. Shaft.
L. Pillars supporting cylinders.

M. Crank shaft bearings.
N. Surface condenser.
O. Reversing gear and lever.
P. Piston crossheads working in slipper guides.
R. Air and Circulating Pumps back of condenser are worked by beams receiving motion from piston crossheads.
V. Valve stems receiving motion from Stephenson link gear.
X. Exhaust pipe leading to condenser.

screw, torsion and strain being correspondingly reduced, allowing lighter construction, or so much additional saving in weight.

If it were possible to devise a perfectly non-conducting substance from which to construct a cylinder, the simple expansive engine would be the best machine so far as profit from expansion itself is concerned, for there is a special loss in the compound, due to expansion of steam while passing from first to second cylinder. This loss is known as the "drop," being the difference between terminal pressure in the high-pressure cylinder and initial pressure in the low. The amount of this drop depends upon details of construction, and is found by comparing "indicator diagrams" taken from both cylinders. Until a perfectly non-conducting lining for cylinders is devised, the loss due to "drop" in the compound is not enough to destroy the benefits derived in other respects.

Summarizing the foregoing we draw the general conclusion that success in economical propulsion by steam is directly promoted by working between the widest possible limits of temperature and devoting the utmost care to providing against losses by dissipation of heat in directions in which it will not contribute to the production of useful power.

The tendency in all attempts at improvements should be toward high pressure, more perfect vacuum, greater expansion, steam jacketing or superheating, non-conducting protectors against waste, and higher piston speeds.

Steam jacketing, which consists of admitting live steam to a special chamber surrounding the cylinder to preserve the heat of the metal of the latter as nearly even as may be, has been found advantageous in engines using saturated steam, but of little or no benefit where the steam is used dry. It may often appear that steam jacketing is of no benefit with the use of low steam, and yet profitable in another engine using high steam, though this is contrary to accepted views among engineers. But the value of jacketing varies with the condition of the steam rather than with its pressure. This will explain the contradictory results observed in practice.

Superheating the steam is preferable. This consists of passing the saturated steam from the boiler through an arrangement of tubes or plates which take up some of the heat escaping into the chimney. Superheated steam is less liable to loss of temperature through condensation in the cylinder.

The arrangement of piping and valves should of course be such as to interfere as little as possible with the free movement of the steam to obviate undue friction and back pressure. In its mechanical aspects, a good engine should have extreme accuracy in the fitting of parts, ample area and durability of bearings and wearing parts, continuous lubrication, freedom from shock and play of parts.

Very much must always depend upon the intelligence and skill of persons charged with the regulating of the machinery. Especially is this true of the compound engine. The key to good management lies in frequent examination of indicator diagrams taken under varying conditions, and from them settling upon the most profitable points of cut-off for each cylinder of the compound if fitted with separate valves and gear. The use of expansion valves on the low-pressure cylinder will not so much alter the total work done by the engines as it will the ratio of work in the two cylinders. The judicious use of these valves, especially at low speeds, will increase the total performance per unit of fuel, but their principal purpose is to regulate the work done in each cylinder. If these valves are not used at low speeds, it will be found that the work done in the low cylinder will hardly be enough to overcome weight and friction, and the low pressure addition may actually prove a drag on the high-pressure cylinder.

The higher the rate of expansion in the low-pressure cylinder, the greater the pressure in the intermediate reservoir and the back pressure in the high-pressure cylinder. Hence, increasing the rate of expansion in the low-pressure cylinder, while developing the power in that cylinder, will correspondingly curtail the power developed in the high pressure cylinder.

If no expansion valves are fitted to the low-pressure cylinder, experiments with varying cut-off can be carried out within limits by means of the links giving motion to the slide valve.

The progress made in the marine engine, from its earliest days up to the present, is simply a record of increased pressure and expansion.

From 5 to 10 lbs. was the usual pressure in the time of Watt. The first engines for screw propulsion used steam at 15 lbs. The mechanism was of the geared kind with jet condensers, and consumed 7 to 10 lbs. of coal per Indicated Horse Power per hour. A little later the direct-acting engine was introduced with steam at 20 ibs., 5 to 6 lbs. being the fuel consumption. Higher pressures and greater expansion followed slowly. Then the surface condenser with its more perfect vacuum than the jet condenser brought down the consumption to 3 and 4 lbs. The boiler pressure rose to 25 lbs. as the troubles from boiler incrustation were partly overcome.

The movement toward still higher steam was kept up until it has reached 60 to 75 lbs. in American river boats, and over 100 lbs. in the non-condensing direct-acting engines of the Western rivers. In the well-known "floating palaces" running on Long Island Sound, the consumption of fuel has been brought down to $2\frac{1}{2}$ lbs. per I. H. P. per hour, but the limit attainable is still far ahead of general achievements in practice. How much room remains for improvement is shown by the exceptional results obtained by special care and design in certain recent constructions, such as the fast torpedo boats and high speed yachts of the period, as well as by new inventions destined in course of time to supersede the present standard compound of two cylinders, just as the latter has displaced the simple expansion machine.

The U. S. Coast Survey steamer Hassler, designed by Mr. C. E. Emery, affords an illustration of the best results yet accomplished in American sea-going service with two cylinder compounds. The Hassler is 151 ft. long, $24\frac{1}{2}$ ft. beam and 10 ft. draft. Engines have cylinders of 18 and 28 in. diameter by 28 in. stroke, indicating 125 H. P., with steam at 75 lbs. pressure. At a speed of 7 knots, the consumption of coal was 1.87 lbs. per horse power per hour.

The following up of the ideas upon which the two cylinder compound has been developed, has led very recently to the introduction of triple and even quadruple expansion machines. In these the total range of expansion is divided between three and four cylinders. The triple expansion is not to be confounded with the three-cylinder compound, in which the exhaust steam from the high-pressure cylinder is allowed to expand simultaneously in two low-pressure cylinders with equal initial pressure. The latter differs from the two-cylinder compound only in arrangement, the work of low pressure being accomplished in two chambers instead of one. In the triple expansion engine, the exhaust is led into a second or "intermediate" cylinder, partly expanded, and its exhaust again into a third cylinder, where the expansion is completed. Despite the multiplication of parts, a material benefit arises from this further subdivision. In the steam yacht Isa, engined by Messrs. Douglas & Grant, of Kircaldy, Scotland, a working pressure of 120 lbs. is maintained, the cylinders being 10, 15 and 18 in. diameter, and 24 in. stroke. They indicate 200 H. P. on 300 lbs. coal per hour, or 1 H. P. for 1½ lbs. of fuel. Observations made upon some of the latest British trading steamers have shown a saving of 19.5 per cent., and an increase of 6 per cent. in speed by the use of triple expansion, and 110 lbs. pressure over ordinary compounds with 90 lbs. pressure, sustaining the conclusions set forth above and pointing out the direction to be pursued for further improvement.

Among the most recent examples in triple-expansion engines, are those of the steam yacht Gladiator, built by Messrs. Ramage & Ferguson of Scotland, the plans of which are taken from *Engineering*. The vessel is 113 ft. keel aforerake, with 20 ft. beam and 13 ft. depth moulded, and is fitted with a Bevis patent feathering propeller, driven by a set of triple-expansion engines. The diameters of the cylinders are 9¼ in., 15 in., and 24½ in., with 18 in. stroke. They are supplied with steam at a working pressure of 150 lbs. per square inch, by a steel cylindrical tubular boiler 8 ft. 6 in. in diameter by 7 ft. 9 in. long, with one of Fox's patent corrugated furnaces. This boiler has a firegrate area of 15 sq. ft., and 500 sq. ft. of heating surface. The particulars of the various pipes shown and numbered in the views are given in the subjoined table.

LIST OF PIPES.

Steam Pipes:

No.		Material	Bore, in.	B. W. G.*
1	Main steam...........................	Copper	2½	8
2	Steam to starting valves..............	"	1	12
3	" donkey...........................	"	1	10
4	" whistle..........................	"	1	10
5	" and water to gauges.............	"	1¼	9
6	" to windlass......................	"	1	12

Exhaust Pipes:

7	Main waste steam.....................	"	4¼	16
8	Bottom blow-off.......................	"	1¾	9
9	Surface...............................	"	1½	9
10	Exhaust from donkey...................	"	1¼	16

Water Pipes:

11	Circulating suctions..................	"	3½	12
12	" discharge.......................	"	3¼	12
13	Air pump "	"	3½	12
14	Bilge discharge overboard.............	"	1½	14
15	Main feed discharge...................	"	1½	9
16	" suction from hot-well...........	"	1¼	14
17	Donkey " sea and bilge........	"	1¼	14
18	" " hot-well.............	"	1¼	14
19	" discharge to three-way cock........	"	1¼	14
20	" " boiler................	"	1¼	9
21	" " on deck...............	"	1¼	14
22	" " to condenser..........	"	1¼	14
23	" " overboard.............	"	1¼	14
24	Fireman's water service...............	"	1¼	16
25	Drain pipe from safety valves.........	"	1	15

Bilge Pipes:

26	Circulating bilge suction.............	Lead	2	
27	Donkey suction from bilge	"	1¼	
28	Suction from aft hold.................	"	1½	
29	" engine-room..................	"	1½	
30	" fore hold....................	"	1½	

* B. W. G.— Birmingham Wire Gauge.

On trial, during a strong breeze and with a considerable sea on, a speed of 8½ knots was attained, the engines indicated 162 horsepower at 145 revolutions. The power being equally distributed between the cylinders being for the

	Horse-Power.
High-pressure cylinder..	54
Intermediate-pressure cylinder................................	53.66
Low-pressure cylinder	54.43
	162.09

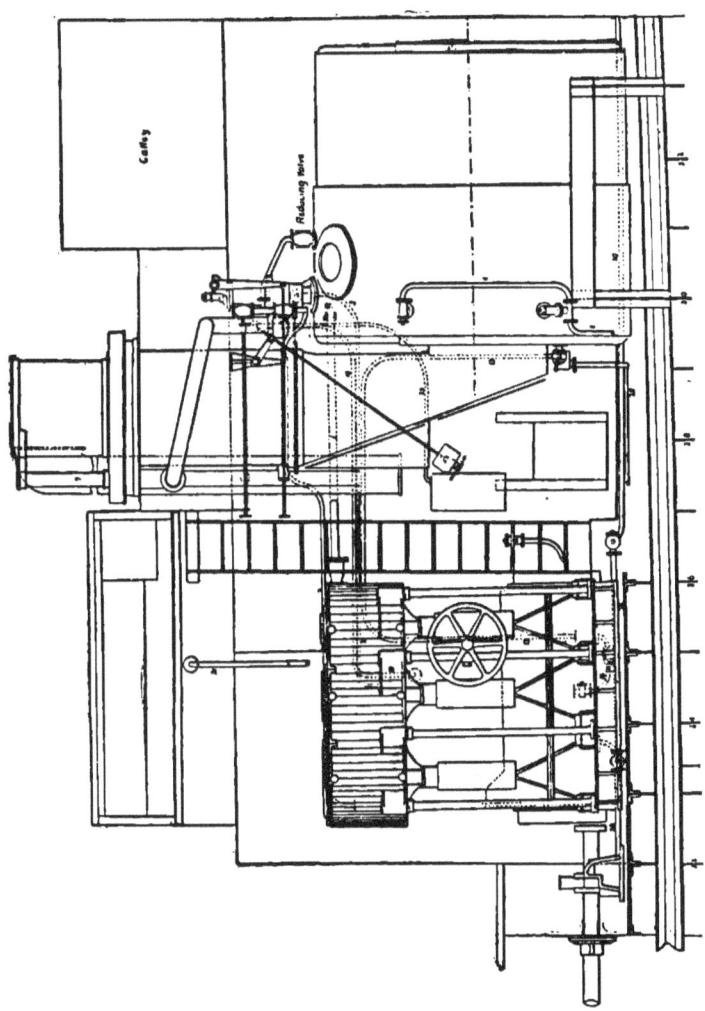

FIG. 5.—ENGINES AND BOILERS OF STEAM YACHT GLADIATOR. FORE AND AFT ELEVATION.

The best general practice of the day has brought us down to something over two pounds of coal per horse power per hour. This is, however, only one-tenth the power derivable from the fuel were all its heat fully utilized, nine-tenths being absolute waste, which never performs service in driving the machinery. The losses can be divided as 70 per cent. of heat rejected in the exhaust steam, 20 per cent. lost by conduction and radiation and the faults of practice, only 10 per cent. remaining for actual utilization. Thirty per cent. of the heat generated in the furnace is lost by passing out at the chimney. Of the remainder which enters the engine 20 per cent. is all of which we can hope to save any portion by improvements in design and construction of the engine itself. In this light, even the best practice is exceedingly faulty and wasteful. Higher results within the 20 per cent. mentioned are to be sought in pressures, expansion rates and piston speeds, of which we have as yet scarcely any conception, unless an entire revolution in the application of the heat generated in the furnace is a discovery destined to radically displace the general procedure with which the engineering world is at present acquainted.

Special instances of great economy, as things go, effect a considerable saving upon more general practice, but even these instances are far from approaching the theoretically possible attainments.

As far back as 1823, Jacob Perkins, a native of Newburyport, Mass., who in later years took up his residence in London, England, experimented with a copper boiler, the sides of which were 3 in. thick, the safety valves being loaded to 550 lbs. per square inch. In 1827 he had attained a working pressure of 800 lbs., but experienced much difficulty in lubricating at the great heat of such high steam, the oils charring and decomposing. Finally this was overcome by substituting rubbing surfaces of a peculiar alloy requiring no lubrication. He also cut off steam at one-eighth the stroke. Pressures of 1400 and even 2000 lbs. were essayed. Concerning these remarkable and vital investigations, Stuart, in describing the work of Perkins, most truthfully comments:

"No other mechanic has done more to illustrate an obscure branch of philosophy by a series of difficult, dangerous and expensive ex-

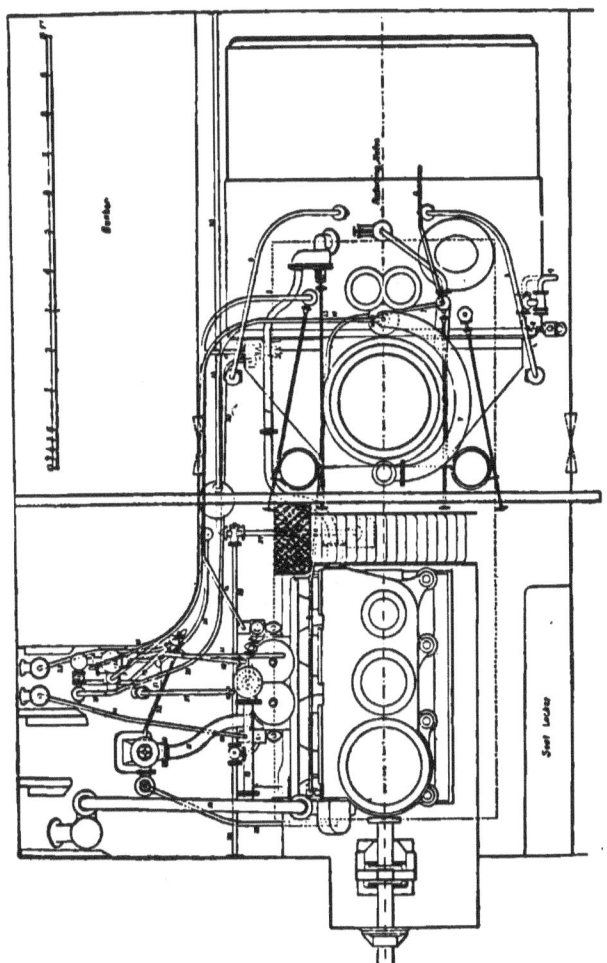

Fig. 6.—Engines and Boiler of Steam Yacht Gladiator. General Bird's-Eye View.

periments; no one's labors have been more deserving of cheering encouragement, and no one has received less. Even in their present state his mechanism bids fair to introduce a new style into the proportions, construction and form of steam machinery."

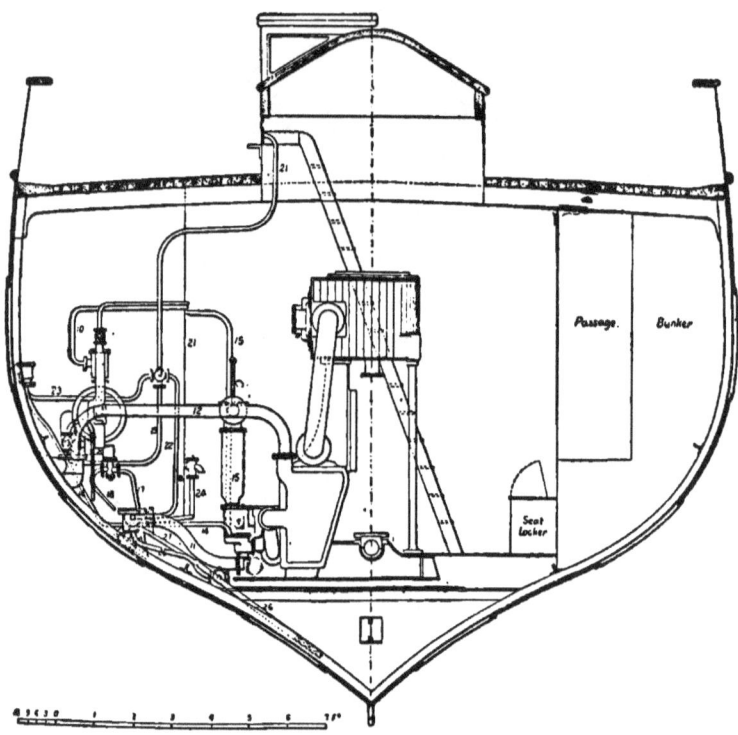

FIG. 7.—GLADIATOR: CROSS SECTION THROUGH ENGINE-ROOM.

That Perkins was far ahead of his time, far ahead even of the present time, and that he was the pioneer of future developments looking forward to something like an approach of practice to theory in steam engine performance, cannot be questioned, but is receiving stronger proof from day to day. When his keen perceptions and zealous labors shall have been fully appreciated, the world will be the great gainer, and possibly will not fail to recognize the genius which piloted at a single bound the fresh advance the engineering

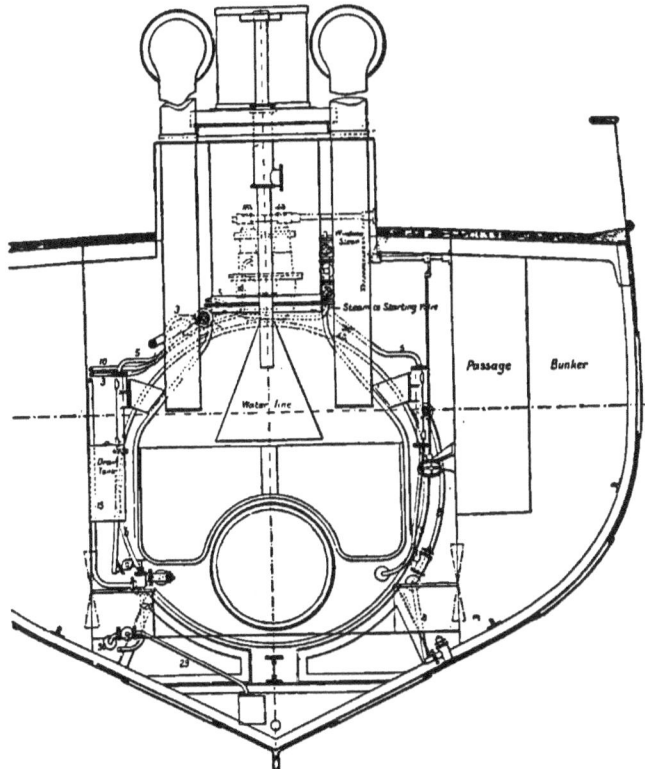

Fig. 8.—Gladiator: Cross Section Through Fire-Room.

world is now tardily following in a modest way. The investigations of the elder Perkins have been pursued by his sons, and through them have compelled at least passing attention and partial recognition, the consumption of coal having been reduced to 1¼ lbs. The gradual introductions of triple and even quadruple compounds and boilers belonging to the "pipe" or "sectional" variety for withstanding great pressures are only so many steps on the road already practically exploited by the Perkins engine and boiler.

The greatest range of temperature is the sum and substance of theoretical engine economy. The rest is a matter of mechanical device for cheaply generating and safely containing high steam and utilizing it with the least loss in the machinery of transmission.

To secure the greatest range of temperature the highest initial pressure must be introduced into the cylinder and the steam allowed to expand to the utmost to part with all its heat energy. But there is a limit in economy to which expansion can be carried, a fact not generally understood or recognized.

The power remaining near the end of a stroke from which steam has been cut off at a very small fraction of the stroke will have declined so rapidly with the simultaneous decrease in temperature that it will be of little service. But as this limit is constantly pushed further with an increase of initial pressure, it follows that the best results will be obtained from the highest pressure to start with, and expansion to the extreme economical limit of that pressure.

The theory of the steam engine is now fairly well understood, and attention is being focussed upon the mechanical appliances with renewed energy and clearer perception, so that we may reasonably expect appreciable advance in saving some of the ninety per cent. of heat energy now wasted, through the introduction of improved or entirely novel methods. When the coal per I. H. P. per hour is measured by ounces instead of pounds, the steam engine will return to us in service a reasonable proportion of the energy stored in nature's great reservoir of combustibles instead of the nominal quantity at our command at present.

II.

BOILER EFFICIENCY.

THE principles involved in designing boilers are few and simple. Although numerous attempts have been made to reach improved results by varying arrangements, the best boilers are nearly all equal in efficiency and by no means superior to some of the earlier types. The modern cylindrical return tubular boiler, now very common in yachts, is not even quite as efficient as the old-fashioned patterns with rectangular fire boxes, long ago abandoned. The cylindrical form has, however, become a necessity with the high pressures now customary, and represents less weight for its power, there being a great saving in bracing and staying over weaker forms.

The requisites in a good boiler are the most thorough combustion of fuel in the fire box without dilution of the products of combustion by excess of air and consequent cooling. This implies as high a temperature as possible in the fire box. Heating surfaces should be so arranged as not to check the draft, and yet take up all the available heat from the gases, which is the temperature in the furnace less the temperature of the gases escaping at the chimney required to produce draft. The grate surface should be sufficient to provide for the consumption of the fuel necessary to develop the heat or power intended.

The mechanical considerations which should govern are strong and cheap construction, with every part accessible for cleaning purposes, no parts weaker than others, and the least opportunity for local corrosion, scaling or burning.

For complete combustion the supply of air should be ample through the grate, and its intermixture with the fuel facilitated. For high temperature in the furnace, the air supply must not exceed that required for perfect combustion. The greater the range of temper-

FIG. 9.—RETURN TUBULAR BOILER WITH DOUBLE FURNACES.

A. Shell of boiler.
B. Furnaces with doors and draft.
C. Ash pits or ash pans.
D. Tubes through which gases return. (The uptake is omitted.)

E. Steam dome for collecting dry steam.
F. Stays to combustion chamber.
G. Manholes for access to interior.
H. Longitudinal stays.

ature between furnace and chimney, the greater the amount of heat available for transfer to the water. If the temperature in the chimney approaches that in the furnace, there will evidently be little heat to give off to the heating surface. Similarly, if the heat in the furnace be cooled down by the admission of too much air, efficiency will be destroyed. The disposition of the heating surface in tubes and plates should not interfere with free circulation of the water in the boiler, and the steam must be withdrawn from the boiler as free from vapor as possible. The cold feed water should enter where the gases are coolest, to prevent needlessly cooling down the heated surfaces.

In practice, the temperature escaping at the chimney has been reduced to 300 deg. Fahr. without checking the draft, and an efficiency of 75 to 80 per cent. of the theoretically possible has been attained. The Clyde boiler is in common use for steam yachts, most frequently with a single furnace. It is of the return tubular variety, and built to the following dimensions:

FIG. 10.—CLYDE BOILER. BUILT BY CHAS. P. WILLARD & CO., CHICAGO.

Length In.	Diam. In.	Diam. Furnace	Tubes No.	Tubes Diam.	Dome Diam.	Dome Height.	Sq. ft. Heating Surface.	Engine.	Weight of Boiler.
42	30	15	40	1½	14	9	42	4½ × 5	775lbs.
44	36	18	56	1½	16	10	62	5 × 7	1100lbs.
48	36	18	56	1½	16	10	80	6 × 8	1200lbs.
50	40	18	70	1½	18	12	90	6½ × 8	1500lbs.
56	44	20	52	2	20	14	105	7 × 8	1850lbs.
68	48	22	68	2	22	16	170	8 × 10	2200lbs.

34   Boiler   Efficiency.

Sectional or pipe boilers are now receiving that attention which is their due in connection with very high pressures, and the question of their generation with safety. It is not too much to expect that the time is not far distant when the shell boiler will be entirely given up

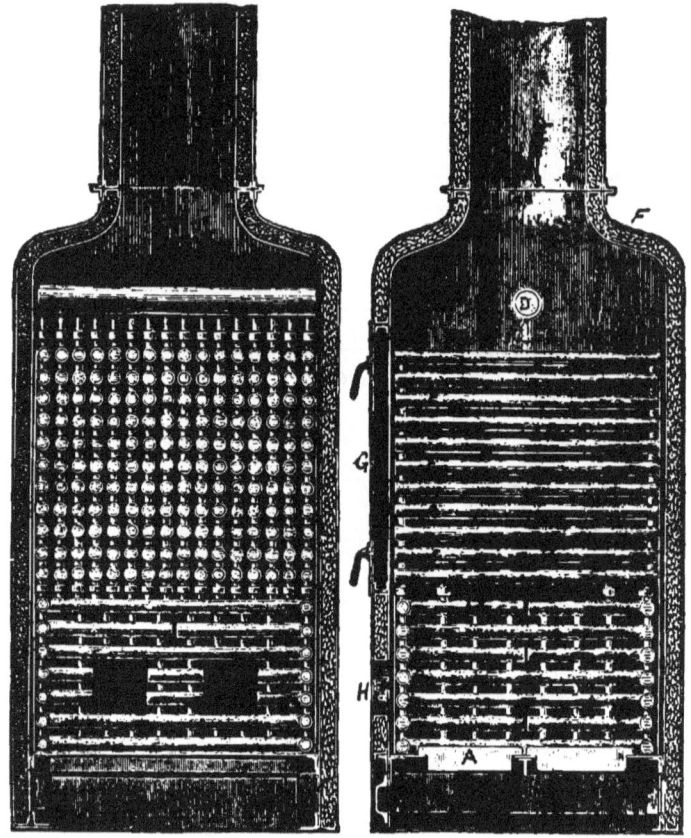

FIG. 11.—THE PERKINS PIPE OR WATER TUBE BOILER.

A. Grate bars.
B. Square furnace built up of tubes.
C. Water-tubes connected by pipe at ends.
D. Steam drum.
E. Smokestack.
F. She tiron casing filled with non-conductor.
G. Door for access to interior.
H. Furnace door.

Boiler Efficiency. 35

and the use of high steam freed from all critical danger. The pipe boiler is an old device, the utility of which was long ago recognized by those experimenting upon high steam. In 1831, Jacob Perkins patented a sectional boiler in which the gratebars were composed of tubes, and subsequently used tubes only, without any large reservoir for water and steam. The economical performance of such boilers with like ratio of heating surface to grate surface is equal to that of the best boilers of the shell type. Their only drawback is the difficulty of keeping up a steady supply of steam, as there is little reserve to draw upon. But this is now being met by special devices.

A pound of pure carbon is theoretically capable of evaporating 15 lbs. of water from boiling point of 212 deg. In practice this is not reached owing to the compulsory loss of temperature passing out of the chimney to produce draft enough to draw air into the furnace. The draft is consequent upon the heat of the escaping gases making them lighter than the air and disturbing balance by the atmosphere at top of chimney and entrance to furnace, the greater pressure at the ashpit door forcing the air up through the grate bars. Part of the combustible gases in the furnace are also carried off without being burnt and some heat is also lost by radiation, owing to imperfect covering of exposed metal surfaces with non-conducting material, such as felt, asbestos or patent preparations. In marine boilers the gasses are allowed to escape at the chimney at nearly 600 deg. Fahr.

The evaporative power of good coal is $13\frac{1}{2}$ compared to 15 lbs. of water for pure carbon. Of the coal burned in steam boilers, thirty per cent. of evaporative power is lost as already explained, and frequently more. Seventy per cent. is all the energy sent to the cylinder less further losses in transmission. Thus actual trial shows that instead of evaporating 15 lbs. of water, the standard for pure carbon, the energy sent to the cylinder per pound of coal is only about $9\frac{1}{2}$ lbs. of water evaporated, and 8 lbs. would oftener be the truth.

Theoretically 12 lbs. of air is enough to consume 1 lb. of coal, but in practice 15 to 25 lbs. are required, owing to imperfect mixing with the fuel and losses through the draft. The product of combustion is carbonic acid gas. Soft or bituminous coal needs more air above the furnace bars than hard coal having a higher per centage of carbon, as the hydrogen liberated from soft coal takes up a larger amount of oxygen in the process of combustion than the same volume of carbon.

A small volume of air is always admitted above the firebars to insure a sufficient supply for the escaping gases. Total area of such openings may be from 3 to 5 in. per square foot of grate. Means must be supplied for regulating the area opened according to the fuel

FIG. 12.—BORDENTOWN "DOGHOUSE" BOILER.

38 in. face ; 38 in. high ; 48 in. long : 50 tubes 2 in. diam. and 41 in. long ; dome 20×20 in.; working pressure 125 lbs.; weight 900 lbs.; engine 6×8 in.

and draft. Too large a supply, as in case of throwing the door open, "dampens" or cools down the fire by absorbing the heat given out from the fuel.

Forced draft has lately received attention. The stoke hole is closed and an artificial pressure produced by means of a "blower" or rapidly revolving fan, promoting combustion and the transmission of a greater amount of heat to the water in the boiler for sustaining a higher pressure or the more rapid supply of steam. Another method is to close up the fire front and blow in air from a fan at the extremity of a tube connecting with the ashpit. As might be expected, the additional power so gained is obtained at corresponding loss in economy, as the process of combustion is forced too rapidly to be thorough, and the loss at the chimney end is increased. With an air pressure in the stoke room equal per square inch to the weight of a column of mercury 6 in. high, the evaporative power of the locomotive boiler of a torpedo boat was found to be about doubled and the consumption of coal likewise.

FIG. 13.—STURTEVANT BLOWER.

In recent steamers built for the merchant service, special arrangements of the furnaces have, however, demonstrated the feasibility of applying forced draft with a view to economy, the fuel being more thoroughly consumed and giving higher results per pound than when relying upon natural draft only. Such prominence has this question of artificial draft assumed of late that we may predict with confidence the universal application of the blast to marine boilers in the near future. Yacht owners in America have the opportunity to lead the way in this as in other improvements instead of following tardily in the wake of the foreign merchant marine.

For 5 sq. ft. of grate surface the diameter of the inlet for the blower should be 5 in. and the outlet 4 in. For 10 ft. of grate, the inlet and outlet are 7½ in. At 2000 revolutions per minute, 1000 cub. ft. of air will be forced into the furnace at an expenditure of ¾ H. P. For 20 sq. ft. of grate, the inlet and outlet are 10½ in., and at 1500 revolutions, 2000 cu. ft. of air will be forced into the furnace at an expenditure of 1½ H. P.

Grate surface is the starting point for apportioning the rest of the boiler. With a given area of grate, a given amount of fuel can be burned, and the utilization of the heat determines the other data of the design. Long grates are difficult to fire, hence the length is usually restricted to 1½ to 2 times the diameter of the furnace. The width of the grate determines the diameter of the furnace in the first place. From 14 to 20 lbs. of coal can be burnt per square foot of grate. The horse-power desired with the presumed efficiency of the boiler enable us to settle upon the area of grate by allowing say one-tenth of a square foot of grate to each I. H. P. The area of heating surface must be from 20 to 30 times that of the grate. For free draft the cross section of the tubes contributing to the heating surface should be about one-seventh that of the grate area. From these considerations the length and number of tubes required is determined after settling upon their diameter. There is much latitude in the proportions given above, and the preferences of builders in constructing boilers of varying arrangement and detail establish the practice in each shop.

Numerous small tubes, while affording more surface, interfere with free draft and are difficult to clean, besides causing violent ebullition and foaming known as "priming" when placed too close. The rule is to give tubes a diameter equal to one-twenty-fourth to one-thirtieth their length, with ¾ to 1¼ inch clearance between them. In short boilers, such as the popular return tubular, in which the furnace is situated under the tubes, a "combustion chamber" is built at the end opposite the door, to further the proper burning of the gases. This chamber is in reality only an extension of the firebox without any grate.

In long boilers of the locomotive type, in which the tubes lead directly from the firebox, such chambers are omitted, as heat enough passes into the tubes to complete combustion in them, and the same applies to vertical tubular boilers. The return tubular boiler is preferred for yachts, owing to the small fore-and-aft space occupied, but the performance of the locomotive boiler is a trifle better, the weight and cost both less, as no combustion chamber has to be built and stayed. For forced draft this type is especially well adapted.

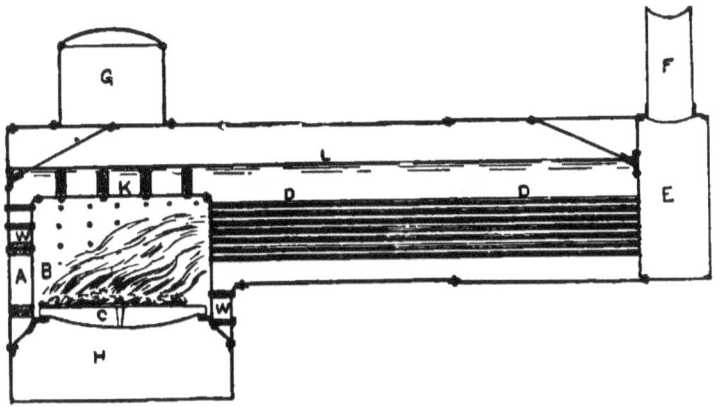

FIG. 14.—LOCOMOTIVE BOILER.

A. Fire door.
B. Furnace or firebox.
C. Grate bars.
D. Tubes.
E. Uptake.
F. Smokestack.
G. Steam dome.
H. Ashpit.
K. Crown sheet of furnace.
L. Water level.
W. Water-legs surrounding firebox.

The crown sheet of the firebox is a very efficient heating surface, but has the drawback of suffering from expansion and contraction and the liability to being left bare of water in a rough sea. "Compensating" stays to the crown sheet and dash plates as well as keeping the crown lowest at the door obviate these troubles. The locomotive boiler has also the advantage of lying low in the vessel. As the size decreases, return tubular boilers rapidly lose in efficiency and are costly and heavy. For small launches, the vertical tubular boiler is in universal use in America. It can be considered as a locomotive boiler set on end, economizing fore-and-aft space.

This style of boiler will be understood from the annexed illustration and dimensions, representing the practice of Chas. P. Willard & Co., of Chicago, Ill.

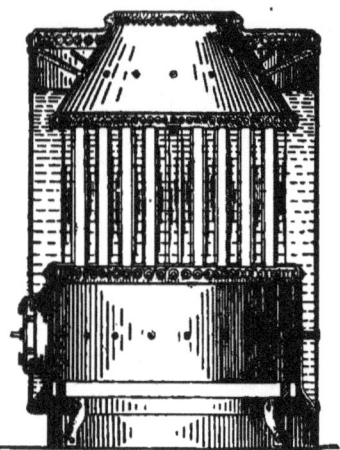

FIG. 15.—VERTICAL TUBULAR LAUNCH BOILER.

It will be noticed that the flues are completely submerged, though in other boilers of the kind they are run up into the head, the gases discharging into an uptake above the boiler shell proper.

Shell.		Flues.			Fire Box.		Sq. ft. Heating Surface.	Engine.	Weight of Boiler.
Diam.	Height.	No.	Diam.	Length.	Height.	Diam.			
22	36	33	1½	12	15	14	20	3 × 4	330lbs.
26	48	43	2	18	18	22	43	4½× 5	580lbs.
30	48	68	2	18	18	26	60	5 × 7	1040lbs.
36	50	87	2	18	18	32	92	6 × 8	1200lbs.
40	53	99	2	18	18	36	100	6½× 8	1530lbs.
42	58	116	2	20	18	38	120	7 × 8	1700lbs.
48	72	160	2	24	24	42	180	8 ×10	2600lbs.

Water tube boilers, in which the heated gases from the furnace pass around the tubes and the water circulates through them, have given comparatively high results, as might be expected from their character. Competitive tests made between the horizontal fire tubular boiler and the Martin vertical water tubular boiler of the U. S. Navy, under like conditions in other respects, have demonstrated the superior evaporative powers of the water tubes. In 1859, a board of U. S. officers reported to the Navy Department that in respect to weight, facility for removing scale, and evaporative efficiency, the water tubular system had an appreciable advantage, but that in the fire tubular arrangement the draft could be forced to a greater

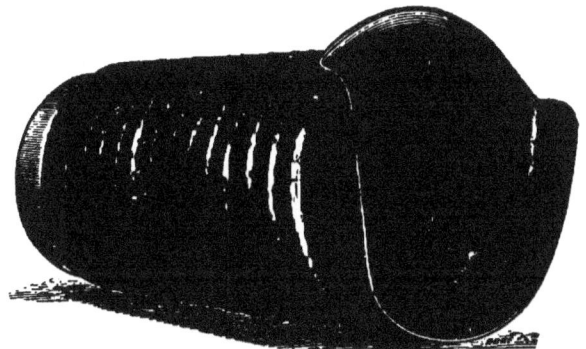

FIG. 16.—FOX'S MILD STEEL CORRUGATED FURNACE FLUE.

extent, of course at a sacrifice in fuel. This report only anticipated the superior efficiency of the more recent styles of pipe boilers, in which there is no longer any difficulty in forcing the draft as desired.

It is always advisable to have ample boiler power. Small boilers for the work must be forced beyond their economical limit, and are short lived.

The number of furnaces is governed by the length and breadth of grate, the diameter of furnaces varying from $2\frac{1}{2}$ to 4 ft. In boilers of 8 ft. length or less, one furnace is the custom. Beyond that two, three being uncommon, except in the largest yachts. Double furnaces have the advantage of enabling successive cleaning and stoking, preserving more even supply of steam than when the door of a single furnace is thrown open. Fox's corrugated mild steel flues are now being introduced in America for furnace construction, giving greater strength and more surface for weight of metal than the smooth plate, and permitting expansion and contraction without stress to the material. A thinner plate, and therefore better conduction, is another point in favor of the Fox flue. The large steam yacht Alva, recently launched by Messrs. Harlan & Hollingsworth, of Wilmington, Del., for Mr. W. K. Vanderbilt, is fitted with four Fox corrugated furnaces to each boiler.

Fire box and combustion chamber must be large, the latter about the size of the furnace in return tubular boilers. The steam space above the water must not be too small, and a large evaporating surface to the water is important, to promote the equal generation of steam without priming. It is customary to place the top row of tubes two-thirds the diameter of the boiler above the bottom. A small body of water promotes rapid steaming, but endangers the tubes and crown sheet from exposure, as the level is more likely to fall too low. Steam drums are attached to boilers for the storage of steam, so that it can be drawn off dry to the cylinder, free from foam.

In respect to circulation, draft, steam space, small bulk of water, ready access for cleaning tubes, lightness, small space occupied, simplicity, facility of repairs, cost, and above all safety, the pipe or sectional type of water tube boilers is superior to the ordinary style of fire tube boilers. The proper estimate now placed upon working with the highest pressure attainable must soon lead to the abandonment of explosive shell boilers carrying large volumes of steam and water and the general introduction of pipe or coil sections in their stead.

In regard to fuel there is equal opportunity for radical improvement. The present dependence upon coal, involving great bunker space, dirt and expense in filling up and in stoking, smoke and hot fire rooms, is a serious drawback to steam yachting. The adaptation of petroleum to the needs of the boiler furnace has in recent years made some progress, though efforts in that direction are still in the first stages. Reports are conflicting and unreliable, but enough has been learned to justify earnest application in the search for practically substituting mineral oils for coal.

Naphtha and electricity have already been made to serve in small launches, the former with great economy in all respects, both sources of power doing away not only with coal as the fuel, but dispensing altogether with the use of steam. Whether we are to see such novel motors successfully devised on a large scale, meeting the needs of seagoing yachts, is for the future to tell. It is certain, however, that evolution toward high pressure pipe boilers and petroleum fuel is the manifest destiny of marine engineering in the yachting fleet as long as steam continues to be the medium for the conveyance of heat energy. There are many yacht-owners whose wealth and position would enable them to serve the engineering world and their own interests by intelligent pursuit of the issues involved, but unfortunately we have so far looked in vain for any appreciation on their part of the field for experiment and profit which lies before them, or of the "moral duty" they owe to the community. The yacht owner in America appears to understand only one, thing, the speed attained by his vessel. Beyond this he seems to be completely divorced from her performances. The organization of the American Yacht Club with the attendant accumulation of data and interchange of experiences will in course of time provide impetus to well directed ambition, and broad profitable results to the engineering world will no doubt displace the petty differences among owners as to who possesses the fastest vessel, regardless of how that speed is produced.

It lies within the power of many a yacht owner to further experimental research and accumulate precise fundamental data which would advance the impending "revolution" in boilers and fuel in one year to a stage it will not reach within one or more decades, if we must continue to rely upon the scattered efforts of individual builders whose interests are as often against innovation as they are in favor of locking up for their private benefit what discoveries they make. The trials carried out upon some of the Herreshoff yachts and torpedo boats are notable instances in exception. They have contributed much to the stock of knowledge concerning pipe boilers, and removed prejudice from the public mind, as the tests were openly conducted under supervision of U. S. officers.

Builders and owners of British steam yachts, who appreciate such results not merely for the maximum rate of speed attainable, but even more for the economics of engine room performance, have also liberally encouraged accurate examination by trial upon individual vessels. Much remains to be done, however, in the way of systematic investigation by thoroughly competent authorities. To such end organizations like the American (steam) Yacht Club now possess the material and wealth, and it rests with them to lift a superficial pastime to the dignity of practical utility. Let us hope that the day is not far distant when the query "What is a steam yacht good for, anyway?" will no longer be heard in the land.

The advance made in British practice should promote higher aims in American steam yachting, for it must be confessed that we are still far behind our cousins abroad. It is true that as a rule our yachts attain a higher maximum speed, but that is because we recognize only that one quality, while British steam yachts are purposely designed for economical and distant cruising without a view to high speed except in special craft constructed solely to that end. While our yachts do not compare favorably with British practice as cruising vessels, they are also inferior in speed to the fast types produced by the chief European nations.

The racing speed of the Atalanta, built in 1883 by Messrs. Cramp & Sons, of Philadelphia, is $19\frac{1}{2}$ land miles, or $16\frac{1}{4}$ knots, and her average sea speed not over 14 knots. The highest speed reached by the Herreshoff boat Stiletto is about 22 land miles or 19 knots, as near as accurate runs have been made. Both of these yachts were designed and engined for high speed vessels in which no limiting considerations were entertained, and their coal consumption is approximately $2\frac{1}{2}$ lbs.

Yet we find abroad that the Spanish torpedo boat El Destructor, built in 1886 by Messrs. J. & G. Thomson, has developed the remarkable speed of 23 knots, or 27 miles, being 8 miles faster than the Atalanta, despite the much greater length of the latter. If it is argued that in small, light constructions like a 100 ft. torpedo boat, the power can bear a much higher ratio to the displacement than in a larger vessel like the 233 ft. Atalanta, we have only to turn to the Cunard liner Etruria to find the speed of the Atalanta greatly exceeded in a large and heavier steamship carrying cargo. It is evident that there is as much room for higher results in speed in American yacht building practice, as there is for distant cruising economy, concerning which scarcely a single example as yet exists in American waters. At a recent trial of a twin screw torpedo boat, built by Messrs. Yarrow, for the Italian Government, a speed of 24.96 knots —equal to 28½ miles per hour was obtained. This is the highest speed ever attained by any description of craft.

The rapid advance made in economy can be gathered from the most recent results, which support the position here taken, that in the maximum pressure and expansion lies the solution of engine economy, and that nothing short of the pipe boiler will meet the demands for the pressures of the near future.

The triple expansion engine now bids fair to displace the compound of John Elder, for it may already be considered as demonstrated that the tripple expansion, with still higher initial pressure, is as much ahead of the compound as the latter was an advance upon the single expansion cylinder worked by comparatively low steam.

The first triple expansion engine was fitted by Messrs. Napier & Sons to the steamer Aberdeen, trading from London to Australia. This was only a few years ago. Last year, almost every steamer sent out from the Clyde was fitted with engines on the same plan, whether vertical or horizontal, for paddle or screw, or even for barges and dredges. Pressure, of course, had to rise in proportion, for reasons previously explained. From the 75 lbs. of a few years ago the demand rose to 100, 120, 140 and 160 lbs., which has become the common working pressure for triple expansion. But even this does not suffice, for Messrs. Denny & Co. have in hand quadruple expansion engines for both paddle and screw steamers, the working pressure to be 180 lbs. Similar engines fitted to the steam yacht Rionnag-na-Mara, 170 ft. long, developed 528 I. H P. at a speed of 12 knots. Steam of 180 lbs. initial pressure was expanded *twelve* times, giving the unequalled economy of 1⅜ lbs. coal consumption per horse power per hour. During a long cruise to the coast of Norway, with steam often down to 145 lbs., the consumption of mixed coal, not of the highest quality, did not exceed 1.45 lbs. Such results are far ahead of the best American practice. The tendency is constantly toward higher pressure and greater expansion, and there is nothing in the latest practice to indicate that the limit has been approached. Bearing in mind the experiments of Jacob Perkins more than half a century ago, and the machinery perfected by his sons and practically tested a quarter of a century ago, the engineering world of to-day is in reality going ahead by catching up with pioneers who have already passed away. Cutting off at half and three-quarter stroke is still common enough. Messrs. Denny & Co. have got it down to one-twelfth, but Loftus Perkins has expanded *thirty-two* times in the engines of the steam yacht Anthracite. It is to be regretted that a fresh experimental start should not have been taken by enlightened yacht owners from the results arrived at by Perkins, instead of merely following in the wake of the merchant service, in which improvements must needs go slow, owing to financial limitations.

III.

BOILER MOUNTINGS.

THE unequal expansion of boiler plates is the chief cause of leakage at the laps and rivets about the bottom of the shell. In raising steam the temperatures at top and bottom differ greatly, and the effort made by the rigid shell to adapt itself to the unequal expansions to suit, leads to strains which may start the seams or even crack the plates and sheer the rivets. With steam at 50 lbs., investigation upon a boiler in use showed a temperature of 300 deg. near the surface of the water and 50 deg. at the bottom, a difference of 250 deg. The heated portion of the boiler was besides several times as large as the cooler portion. The amount to which an 18 ft. plate will expand at 250 deg. is ⅜ in. In this proportion, the top of the boiler was lengthened, while the bottom scarcely expanded at all, hence it is easy to understand the probable strain of such unequal treatment. The only way to avoid extreme tension, is to promote circulation of the boiler water as much as possible. For this purpose the hydrokineter has been introduced in England with marked success in the yachts of the large class. This attachment is seldom to be met with in American practice, but the fact that by its means the temperatures at the top and bottom of the boiler have been kept within 20 deg. of one another, shows its efficiency. This instrument consists of three nozzles capping one another, the first having a flange fitted inside on the bottom of the boiler shell. Steam from the donkey or winch boiler is admitted by a valve, and as it rushes through the hydrokineter, the water in the bottom of the boiler is drawn in through annular openings between the nozzles and propelled as a jet out of the third nozzle with considerable force, thereby setting up circulation in the boiler water.

The attachments of a boiler comprise: Safety valve, steam gauge, water cocks, water gauge, salinometer, feed valves, blow valves, dry pipes and stop valves. Uptake, dry combustion chamber, blast, damper, whistle, injector and funnel may also be deemed attachments. In yacht boilers the combustion chamber is always an integral part of the construction. The accompanying diagrams will elucidate the parts in detail.

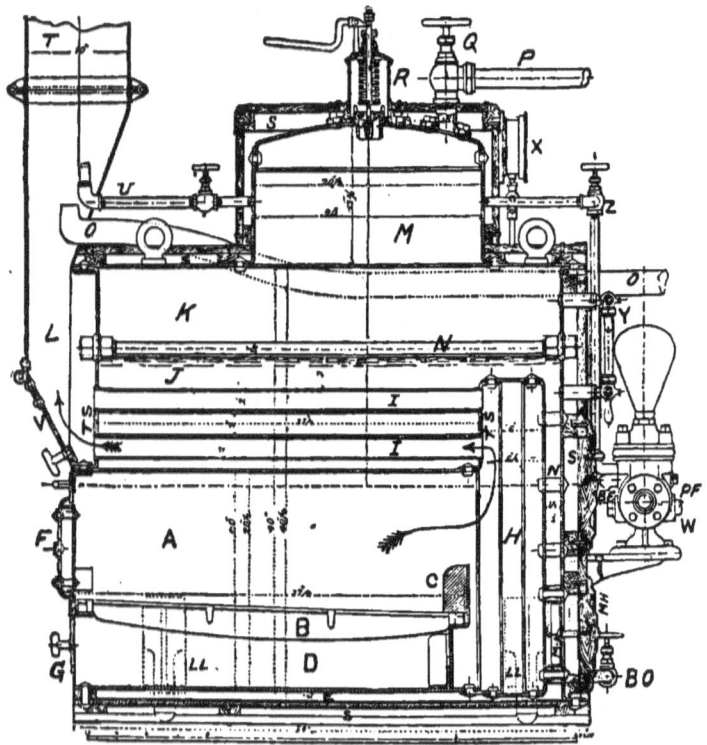

FIG. 17.—RETURN TUBULAR BOILER, U. S. NAVY LAUNCH.

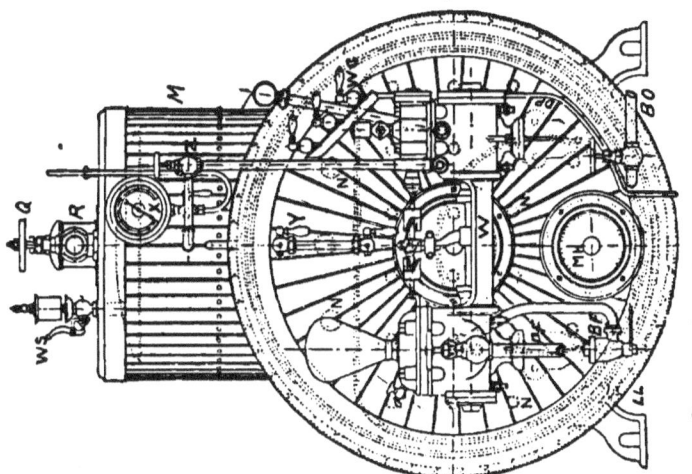

Fig. 19.—Rear View with Feed Pump.

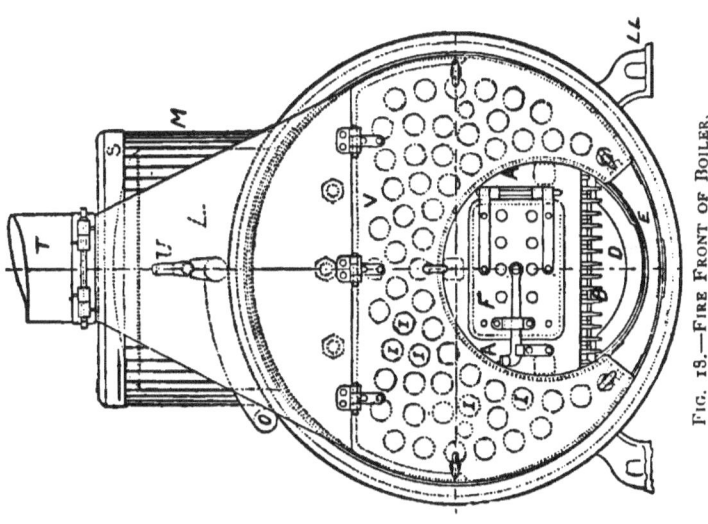

Fig. 18.—Fire Front of Boiler.

[49]

EXPLANATION OF DIAGRAMS.

A. Cylindrical furnace, within the cylindrical shell of the boiler. It is supplied with a "fire front," containing the furnace door, F, and ash pit door, G. It is riveted to the tube sheets, T S, in front and rear, and is completely surrounded by water circulating under the ash pit through the space E.
B. Grate bars, with fire bridge, C, at back, to retain the fuel. This bridge is lined with fire brick, shown by the diagonal shading.
D. The ash pit.
E. Water space underneath the ash pit.
F. Furnace door, supplied with lining sheet on the interior and holes for draft.
G. Ash pit door or damper.
H. Combustion chamber, built at back of furnace. It is stay-bolted to back of outer shell as at N. The gases from the furnace pass into this chamber for further mixture and combustion, and thence return to the front through the tubes I, as indicated by the arrow.
I. Fire tubes surrounding the furnace. They are "expanded" into the tube sheets at each end, the front ends leading into the uptake, L, thence into the smokestack, T.
J. Water space above the furnace.
K. Steam space above the water level.
L. Uptake.
M. Steam drum or dome, flanged and riveted to the boiler shell. The latter is pierced with holes to allow the steam access to the drum, from which the steam is taken free from foam by the steam pipe P on top of the drum. The dome and perforations take the place of "dry pipes" in large boilers without drum.
N. Tie rods between ends of boiler shell, called stays.
O. Exhaust pipe from non-condensing engine, leading into smokestack, T.
P. Steam pipe for supplying the engine.
Q. Stop valve for regulating and shutting off steam supply to engine.
R. Safety valve, depending upon a spring for its resistance to pressure from below.
S. Spaces for non-conducting covering between boiler shell and external wood lagging in which the boiler is encased.
T. Smokestack, with joint for lowering.
U. Steam blast from drum to smokestack to increase the draft.
V. Door to uptake, by which all the tubes can be got at for cleaning.
W. Independent pump for feeding the boiler.
X. Steam gauge, for indicating pressure within the boiler.
Y. Glass water gauge, showing height of water in boiler.
Z. Steam valve to steam end of independent feed pump.
P F. Feed suction, for supplying water to the pump.
B F. Boiler feed pipe from pump.
B O. Bottom blow-off, by which the water in boiler can be blown out. Also for filling boiler when cold, owing to higher level of sea at boat's side.
L L. Cast iron legs to boiler. These legs are bolted down to the boiler keelsons.
M H. Hand hole plates, which can be removed to clean out combustion chamber.
W S. Steam whistle in top of drum.
W C. Water cocks, for testing height of water level.
D P. Drip pipe from water cocks, leading into bilge of boat.

Boiler Mountings. 51

The bolts shown on top of boiler are for hoisting in and out of boat. This boiler is suitable for a launch 33 ft. long, 8 ft. beam and 3 ft. 9 in. draft. The shell is 3 ft. 4½ in. diameter outside ; furnace 20½ in. external diameter, the plates of shell and furnace being ¼ in. thick, as also the tube sheets. There are 60 return tubes 3 ft. 1½ in. long and 2 in. diameter outside. Area of grate 4.5 sq. ft., heating surface 125 sq. ft. The dome is 20 in. diameter, 14 in. high. Weight complete 2350 lbs.

SMOKESTACKS are proportioned to the cross-sectional area of tubes and should gradually contract to meet the cooling of the gases, although in practice the contraction is confined to the uptake. Friction would limit the height of a chimney, but the fault is always the other way, most stacks being too short for good draft

SAFETY VALVES are made in great variety, differing in details of design, but all depend upon the same principles. The common safety

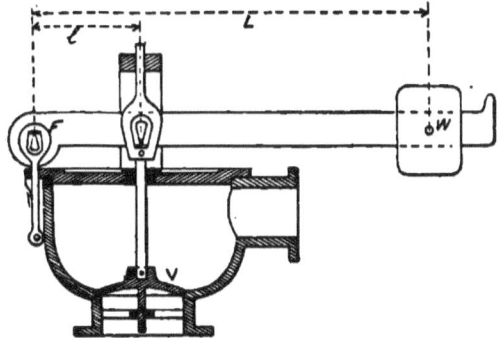

FIG. 20.—COMMON SAFETY VALVE.

valve is shown in the diagram. For loading the valve lever the measurements are made from the fulcrum F. The long arm is the distance from F to center of weight W, and the short arm is the distance from F to vertical center-line of the valve-stem. We have from the principles of mechanics the formula $P l = W L$, in which P is the desired pressure under the valve and W the weight hung upon the lever. This would not take into consideration the weight of the valve, its stem and the lever, for which allowance must be made in fixing upon W or its distance from the fulcrum F. By removing W and attaching a spring balance at end of lever the weight of the gear itself can be obtained, which must be subtracted.

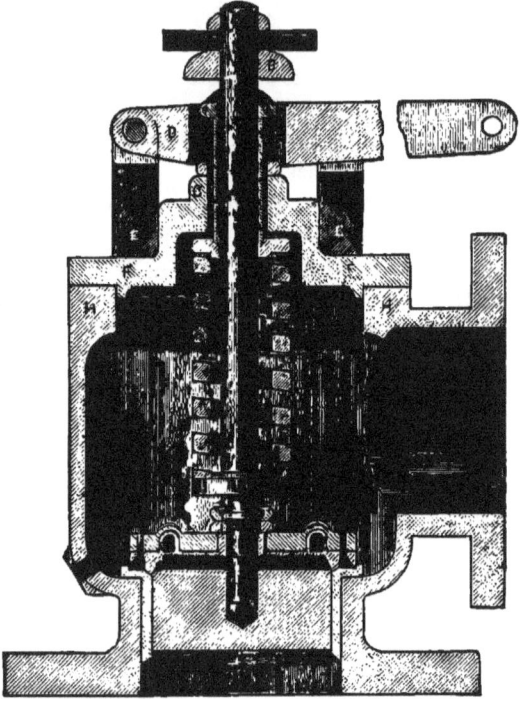

FIG. 21.—MARINE POP SAFETY VALVE. AMERICAN STEAM GAUGE CO.

Example: A safety valve is 6 in. diameter and it is required to blow off at 60 lbs. pressure per square inch; the weight W is 175 lbs., and that of the gear 120 lbs. The short arm l is 4 in. Required the distance from fulcrum at which W must be suspended. The area of a 6 in. circle is 28.26 sq. in., then

$$L = \frac{[(28.26 \times 60) - 120] \times 4}{175} = 36 \text{ in.}$$

In small safety valves, springs take the place of the weight. The strength of the spring forcing the valve into its seat can be regulated by screwing a nut on the stem up or down. The pressure at which it blows off will be shown by the steam gauge. There should be a limit to the compression of the spring by the nut, to prevent "bottling up" the steam beyond the regular working pressure for which the

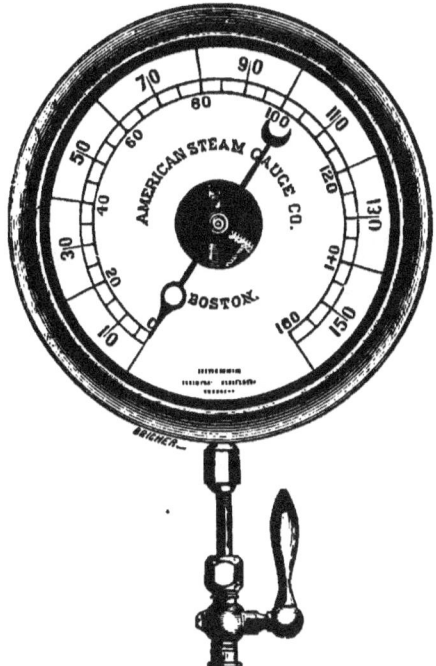

FIG. 22.—BOURDON STEAM GAUGE.

boiler is intended. The area of the valve should not be much less than half a square inch per square foot of grate, and the "lift" should be equal at least to one-quarter the diameter. The escape pipe should have an area equal to that of the valve. Spring valves are preferable in small vessels, as the weight on the lever variety is affected by the rolling of the vessel. In construction the valve should be simple and strong, so as not to clog or have free passage of steam obstructed by details of construction   Periodical inspection of the boiler will necessitate modifying the safety valve to suit the increasing age of the boiler. Explosions are frequently to be ascribed to carrying the original pressure on old and more or less worn out boilers.

STEAM GAUGES in general use on board of yachts depend for their action upon the expansion of thin metal tubes or diaphrams communicating motion to the index arm through a system of levers. In

BOURDON STEAM GAUGE—INSIDE.

the Bourdon, for example, the steam is admitted through a pipe connected to a circular tube of oval section. As pressure increases, the tendency will be to straighten out the tube. This will cause the index arrow to revolve in obedience to the intervening arc and pinion. The dial is marked off by the manufacturer from comparison with a standard mercury gauge. Another form, such as the Utica steam gauge, derives its action from the expansion of light metal plates in a casing or drum, the corrugated sides of which answer vertically to increase of pressure. The motion thus obtained can be multiplied by a system of levers.

Boiler Mountings.

The steam pipe connecting with the boiler is made with a neck or syphon to collect water which will act as a cushion between the live steam and the working parts of the gauge. A cock at the top of the bend prevents accumulating too much water by occasional attention. This whole arrangement is shown at X in Fig. 19. These gauges are sensitive and require care. They should be cleaned from time to time, else the thickening of the lubricants or stoppage of passages may totally deceive the engineer. If supplied with a blow-off cock on top, the steam can be turned off from the boiler and the cock opened. The gauge should then run back to o. If it

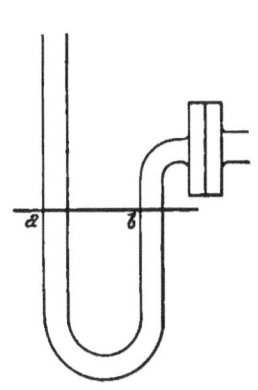
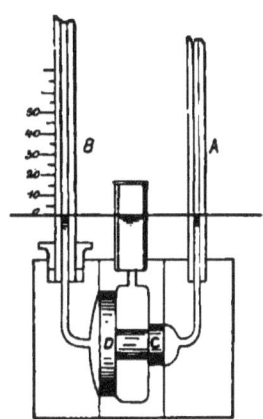

FIG. 23.—MERCURY GAUGE.   FIG. 24.—IMPROVED MERCURY GAUGE.

does not, the number of pounds still indicated should be subtracted from the total for the true pressure. Cocks should be opened gradually to prevent shock to the mechanism.

STANDARD MERCURY GAUGES are also fitted to the larger class of yachts. They depend for their action upon the weight of a column of mercury contained in a glass tube. The simplest form has a cross area of 1 sq. in. The end b is connected with the boiler and the end a left open, the neck being filled with mercury to some level $a\,b$. The weight of 2 in. of such a column equals 1 lb. Hence,

56   *Boiler Mountings.*

if the steam presses down the mercury 1 in. at *b*, it will rise 1 in. in the tube *a*, a difference of 2 in. in the level. A scale of inches attached to *a* will indicate the number of pounds pressure in boiler. The area of the tube need not be 1 in., but can be reduced to any fraction, providing the area exposed to the steam is reduced in the

FIG. 25.—SAFETY COMBINATION GAUGE. AMERICAN STEAM GAUGE CO.

same degree. A small tube prevents oscillation of the mercury. The length of tube *a* would become impracticable for high pressures, and must be kept down by the intervention of compensating pistons.

In Fig. 24 is shown the open top gauge devised by Chief-Engineer A. S. Greene, U. S. Navy. In a casting made up of layers a piston is introduced with the small area C exposed to the steam end and the large area D to the scale end. A small descent in the tube A counterbalances a large steam pressure owing to the excess in area of piston C over that of tube A, and a still smaller ascent takes place in the scale tube B, owing to excess of piston D over C. In this way the scale tube is kept down to convenient height.

WATER GAUGES are tapped at proper heights into the boiler shell and secured with nuts on inside. They indicate the water-level and comprise a glass measuring tube and a series of three cocks, both sets being imperative as a check upon one another, for low water is a very serious matter and the most prolific cause of explosions. If the level were to fall below the tubes or crown sheet of furnace, the iron would quickly be brought up to red heat and then burnt, or warp permanently out of shape. If the feed be speeded up, the sudden contact of cold water with the hot iron causes the water to flash into steam, rending the boiler to pieces. The glass gauge is fitted with a stop cock at each end, so that it can be shut off from the boiler, and a broken tube replaced. There is also a blow-off cock at the bottom for drawing off the water occasionally so that upon refilling from the boiler it may be noted whether the glass measures uniformly and is not stopped up. In dark stoke rooms, a light should be on hand for correctly noting the level. The water cocks are three in number, so placed that the upper one should show steam, the center cock steam and water from about water level, and the lower one always water only. It is the duty of the engineer to examine by these so frequently that the level shall never descend below the third cock.

FUSIBLE PLUGS are further safeguards against low water, and ought to be fitted to all boilers. The plug consists of a hollow brass socket screwed into the crown sheet of the furnace. A smaller socket is screwed into the fixed one and a hole through the small one filled with an alloy of lead and copper which melts at comparatively low temperature. When the water level falls below the level of the plug, the alloy melts and the steam escapes into the furnace, giving warning and dampening the fire. A spare plug can then be inserted after cooling down and the boiler refilled. The plug needs watching to prevent covering by scale and should be renewed occasionally, as a hard skin forms on the alloy with age.

Another safeguard against low water is the alarm whistle. This is attached to the boiler at lowest water level. If the water is allowed to drop below that, the steam rushes into the whistle and gives the alarm. It is specially to be commended for boilers having little water space or depth above the crown sheet of the furnace.

FIG. 26.
LANE'S LOW WATER AND ALARM GAUGE.

Put the line on cylinder exactly on a level with the waterline in boiler; screw the three-way valve into boiler the same distance below waterline in boiler that it is from the line on cylinder to center of three-way valve when screwed in its place to bottom of cylinder; then couple cylinder on to valve, and connect pipe where you get dry steam at top. Steam should be blown through the three-way valve once a week. When the three-way valve is screwed out, it allows the passage of water from boiler through up into cylinder, and then the gauge is in working order. When it is screwed in, it allows the steam to pass down through the cylinder, while at the same time it shuts the water from the boiler. Place the cylinder as near the boiler as possible.

THE FEED WATER passes into the boiler through a check valve in which the valve has no stem, but constantly drops back into its seat, thereby preventing the boiler water from escaping back into the feed pipe. Each stroke of the feed pump forces the valve up, so that the feed is delivered intermittently. Applying the ear to the valve will tell whether the check is working properly. If the valve becomes heated it is a sign that back water from the boiler is escaping. A

Boiler Mountings.

detached stem serves in larger vessels to regulate the feed by screwing up or down, varying the lift of the loose valve. It is better, however, to regulate the feed by a separate valve, allowing the check to have free play to keep it clear. An escape valve is sometimes fitted to the feed pipe to prevent overloading it, risking a burst in the pipe.

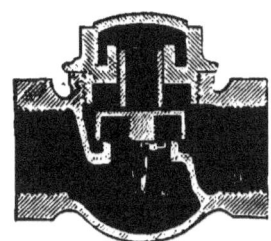

FIG. 27.—CHECK VALVE. JENKINS BROS., NEW YORK.

Launches and small yachts are now commonly supplied with an "injector" instead of feed pump. There are numerous devices of the kind in the market, the original invention being known as the Giffard. The injector is more reliable than a small pump, as there are no valves to clog and fewer parts. It occupies less space and requires less attention. The steam used in its operation is condensed in the feed water, heating it and returning it to the boiler in that state without a special feed heater. The working of an injector can be explained as follows : Steam from the boiler enters a conical receiving tube at the head of the instrument. At the mouth of the cone the escape of the steam will suck the air from a branch pipe leading to the sea through the boat's bottom, and the sea water or suction will be drawn up the branch pipe owing to the partial vacuum created. The suction water then meets the steam at the nozzle of the receiving cone in the instrument. The water receives the impact of the steam, condenses it and acquires considerable velocity as a jet, enough to force its way through a delivering tube, past a check valve, into the boiler against the boiler pressure. The power of the feed to enter the boiler is derived from its weight moving at the velocity acquired from impact with the steam in the injector. If too much water be admitted, it will not acquire energy enough, and if

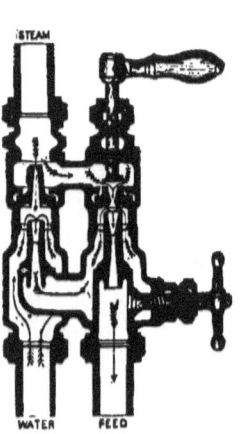
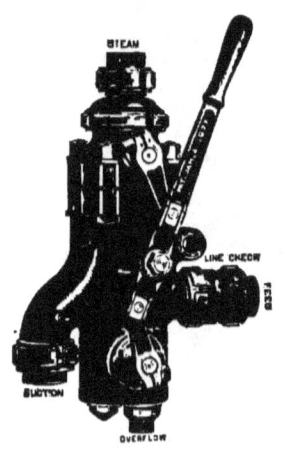

FIG. 28. FIG. 29.
HANCOCK INSPIRATOR. HANCOCK LOCOMOTIVE INSPIRATOR.

too little in proportion to the steam pressure, the energy will also be diminished for lack of sufficient weight, as energy is the product of weight and the square of its velocity. Hence an injector with fixed parts would be suitable only to a given steam pressure. For different pressures, the amount of water admitted would need regulating by a valve or automatic piston arrangement. The first would need constant attention, and the latter is liable to fail in its action. To overcome these drawbacks of the ordinary injector, the Hancock Inspirator was devised and is now in general use in launches and yachts. Fig. 28 shows the instrument in section, and Fig. 29 is an elevation of the locomotive pattern adapted to marine use, and operated by one handle. The arrows in Fig. 28 make the action clear. There are no pistons or movable parts. An ordinary injector is a single apparatus requiring adjustment for different pressures, but the Hancock injector is double, one-half being a "lifter" and the other the "forcer," the first raising and the latter delivering the feed at any pressure. It will lift water 25 ft. with about 45 lbs. of steam, and will take it 140 deg. Fahr. on a lift of 3 or 4 ft. On a lift of 25 ft. the suction should not be over 100 deg. The temperature of the delivery will be increased about 100 deg. The essential conditions of its successful operation are a tight suction and dry steam direct from boiler, not from another pipe.

The suction pipe should be two or three sizes larger than the connection for a high lift or long draft. In attaching, be careful to blow out steam pipes before connecting to clear them of red lead or filings which would stop up the instrument. The suitable size is determined by the capacity of the boiler, as follows: When consumption of fuel is known, the pounds of coal consumed per hour will be the number of gallons evaporated per hour and to be supplied by the inspirator. When the grate surface is known, and the draft natural, multiply the grate in sq. ft. by 9 for the gallons evaporated per hour. For forced draft allow 50 per cent. more. Check valve to boiler should never be smaller than $\frac{3}{4}$ in., the usual $\frac{1}{2}$ in. valve being worthless. In light draft yachts subject to rolling or pitching in a sea, the suction is prevented from drawing in air by the addition of a small well in which the sea pipe rises inside of an elbow above the level of the inspirator suction, thereby keeping the elbow or T full of water as a stop to the suction, allowing air taken in to escape at the top of the well. To start the inspirator draw back the lever sufficiently to open the overflow valves and bring water, and then draw back to the stop, which closes the overflow, and diverts the jet into the feed pipe leading to the boiler.

BLOW-OFF VALVES are fitted to all boilers, one near the level at which the water is carried, known as the surface-blow, and the other at the bottom, known as the bottom-blow. As the boiler water is evaporated into steam, the remaining water becomes denser, for the salt of sea water and other impurities are left behind. A deposit, principally sulphate of lime, will be precipitated upon the interior surface of the boiler shell and tubes, forming a thin sheet of scale which is a very poor conductor of heat. If this scale accumulates and hardens, serious damage will result from overheating the surfaces exposed to the fire, or "burning," as the cooling boiler water on the inside does not come in contact with the metal for its protection.

To prevent such results, a portion of the boiler water is periodically "blown off" and replaced by additional feed water, so that the average density of the water in the boiler may be kept below the point at which serious deposit takes place. The water blown off has already been heated while the fresh feed is taken in cold, hence a waste of heat necessarily accompanies the operation. This is called the "loss by blowing off." In part this is made up by passing the cold feed through special feed water heaters deriving their heat from the exhaust steam of non-condensing engines. In condensing engines, however, the only source of heat available is from the water blown off. This is led through a series of pipes surrounded by the cold feed, so that part of the escaping heat is transferred to the fresh feed.

The higher the temperature of the water, the greater will be the precipitation, hence it is the custom to provide a continual surface discharge termed "surfacing" or "scumming." The density least favorable to the formation of scale is a matter of judgment and experiment. The common rule in the U. S. Navy is to carry the water at a density of $1\frac{3}{4}$ by the salinometer, and most engineers follow this direction. Mr. Geo. W. Baird, U. S. N., has found a higher density preferable. The ground taken was that if we carry the water at a high density, we do not have to blow off so much nor supply so much sea-water feed. If we blow less, we pass less seawater through the boiler and less scale is accumulated. The solid matter other than the sulphate of lime in the boiler does no harm until the density of about $3\frac{1}{2}$ is reached, which will be the limit beyond which we cannot go. Experiments were carried out on two naval vessels a few years ago to test Mr. Baird's proposition. In

Boiler Mountings.

one boiler the water was kept at the regulation density of 1¼, and in the other at 2½, accompanying conditions being in other respects alike. After a four months' cruise, it was found that the boiler carrying water at the low density of 1¼ was cleaned of scale twice, and that the boiler carrying the water at the higher density of 2½ required no "scaling" at all, and at the end of the cruise was freer from scale than the other.

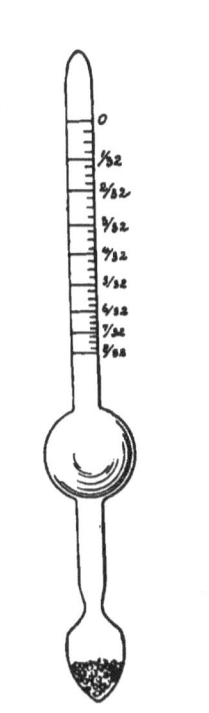

FIG. 30.—SALINOMETER.   FIG. 31.—LONG'S SALINOMETER. AMERICAN STEAM GAUGE CO.

THE SALINOMETER consists of a glass tube with ballast in a bulb at the lower extremity. When dropped into fresh water the line of flotation supplies the o of the scale. If placed in water containing 1 lb. of saline matter in 32 lbs. of water, it will rise to $\frac{1}{32}$ of the scale; when in water containing 2 lbs. of saline matter in 32 lbs. of water, it rises to the $\frac{2}{32}$ mark on the scale, and so on. If the water in a boiler is said to have a density of $1\frac{1}{2}$, it means that the salinometer would float half way between $\frac{1}{32}$ and $\frac{2}{32}$ of the scale. A density of $2\frac{1}{4}$ means that the instrument will float one-quarter the distance between $\frac{2}{32}$ and $\frac{3}{32}$. The scale steadily contracts down the instrument, because at greater densities the displacement of the instrument becomes less and less, while its own weight remains constant. The salinometer is graduated for a temperature of 200 deg. Fahr. and will not serve for other temperatures, as the density varies with the latter. A difference of 10 deg. in the water will cause a difference of $\frac{1}{8}$ in the scale. To facilitate the use of the instrument in practice, the boiler water, after being drawn off, must be cooled down to 200 deg. by some method which will allow continuous observation. This is accomplished by passing the boiler water through a coil surrounded by cold water according to the plan of Chief Engineer Fithian, U. S N., or by having a separate cylindrical casing to receive the hot water and steam from the boiler, the water flowing over into the tube in which the salinometer floats, while the steam escapes without reaching the observer. The salinometer tube is surrounded by cold water in an outer tube. By regulating the supply of boiler and cooling water, an even temperature of 200 deg. can be maintained. This is the plan devised by Sewell and Long of the U. S. Navy.

Such instruments are supplied only to sea-going yachts. In launches and steamers which make a harbor frequently, the salinometer is seldom brought into use. The boiler water is entirely changed after every run and the engineer blows off occasionally according to his judgment. In condensing engines the frequent renewal of all the boiler water is an advantage, as it gets rid of deleterious matters from the condenser which the salinometer would not detect. By filling up with fresh water before starting, a day's run can be made before the water will reach a density of 3, even if the feed be sea-water.

Boiler Mountings. 65

FEED PUMPS of the direct-acting variety with boiler connections supply feed water when an injector is not used, and are always a part of sea-going machinery. They work independently of the engine, although in launches frequently from an eccentric on the shaft or from the piston crosshead. The independent pump can be run after the yacht has come to an anchor or is in dock, so that the steam can be run down and the boiler filled ready for firing again.

FIG. 32.—WORTHINGTON BOILER FEED PUMP.

The Worthington Feed Pump, like all others made by the Henry R. Worthington Hydraulic Works of Brooklyn, N. Y., are of the "duplex pattern." There are two steam and two water pistons, as the perspective illustration will show. Such arrangement insures smoothness of working, efficiency and reliability. The valves of the steam cylinders are ordinary slides, the simplest and most reliable of all kinds. They receive motion from a vibrating arm which swings through the whole length of the stroke free from sudden blow. The work of the pump being divided between two engines, the wear is also divided and the lifetime increased. The discharge is also continuous and steady.

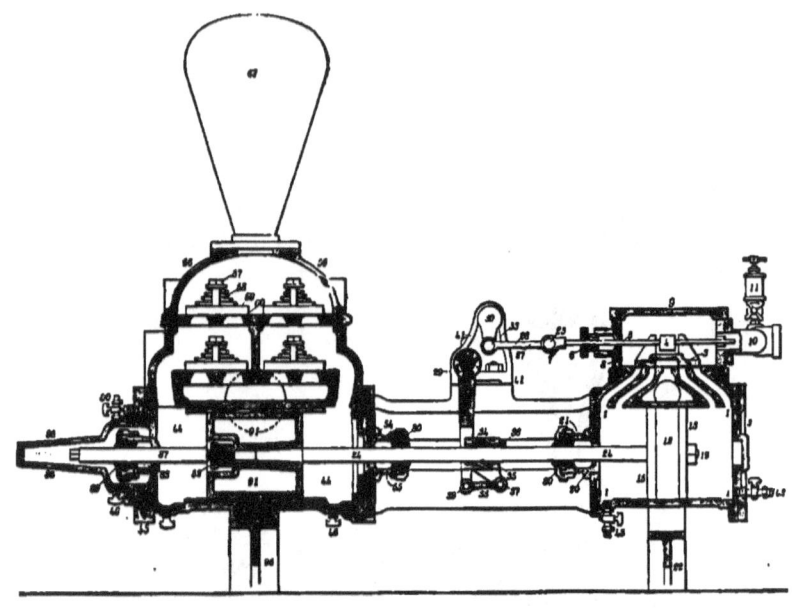

Fig. 33.

NOMENCLATURE OF THE WORTHINGTON FEED PUMP.

1. Steam cylinder (No. 1 or No. 2 side).
2. Steam cylinder head.
3. Slide valve.
4. Valve rod nut.
5. Valve rod.
6. Valve rod gland.
7. Valve rod head.
8. Steam chest.
9. Steam chest cover.
10. Steam pipe.
11. Lubricator.
12. Piston ring.
13. Piston follower.
15. Piston body.
16. Piston spring.
17. Piston tongue.
18. Piston tongue spring.
19. Piston nut.
20. Piston rod stuffing box.
21. Piston rod stuffing box gland.
22. Steam cylinder foot.
24. Piston rod.
25. Valve rod head pin.
26. Long valve rod link.
27. Short valve rod link.
28. Long lever.
29. Rock shaft key.
30. Upper rock shaft.
31. Lower rock shaft.
32. Short lever.
33. Crank pin.
34. Crosshead.
35. Crosshead position pin.
36. Crosshead key.
37. Crosshead pin.
38. Crosshead link.
39. Lever pin.
41. Cross-stand.
42. Waste cock.
43. Pet cock.
44. Water cylinder.
45. Water cylinder head.
46. Thumb plugs.
52. Water piston nut.
53. Water piston jam plug.
54. Plunger rod stuffing box.
55. Plunger rod stuffing box gland.
56. Force chamber.
57. Valve guard.
58. Valve spring.
59. Valve.
60. Valve seat.
67. Air chamber.
80. Stuffing box follower.
81. Sectional stuffing box follower.
82. *Side-feed crosshead.
83. Side-feed plunger.
84. Side-feed stuffing box gland.
85. Side-feed pump barrel.
86. End-feed pump barrel.
87. End-feed plunger.
88. End-feed stuffing box gland.
89. End-feed stuffing box follower.
90. End-feed air cock.
91. Solid water piston.
92. Packed water piston body.
93. Packed water piston follower.
94. Packed water piston packing.
95. Packed water piston setting-out ring.
96. Water end foot.

* Side-feeds were attached to pumps built prior to 1885. The End-feed is now standard.

Boiler Mountings.

The capacity of the pump is shown in the following table of sizes, the number of strokes per minute being from 75 to 150.

Diameter of Steam Cylinders.	Diameter of Water Plungers.	Length of Stroke.	Gallons delivered per minute by BOTH plungers at stated number of strokes.	Diameter of plunger required in any single cylinder pump to do the same work at same speed.
3	2	3	6 to 16	2⅞ ins.
4½	2¾	4	15 to 30	4 ins.
5¼	3½	5	30 to 60	5 ins.
6	4	6	50 to 80	5⅝ ins.

In small sizes for steam launches, Chas. P. Willard & Co., of Chicago, use a single cylinder pump correponding to the following dimensions:

Size Steam Cylinder.	Size Water Cylinder.	Length of Stroke.	Capacity Per Minute.
2½ inches.	1½ inches.	3 inches.	3½ galls.
3 inches.	1¾ inches.	3 inches.	4½ galls.
3½ inches.	2¼ inches.	3 inches.	7¾ galls.

FIG. 34.—FEED PUMP FOR LAUNCHES.

The pump is driven up to about 150 strokes per minute. Injectors should always be added, so that if the pump gives out, there will be no danger of low water in the boiler. Independent steam pumps are more economical than injectors.

Long straight steam pipes should be avoided, as they will break in consequence of expansion, especially where high steam is used. Elbows should be introduced to give the needed play or else special

FIG. 35.—FEED PUMP WITH PLUNGER.

"expansion joints" must be fitted. The working parts of such joints should be of brass or lined with the same to prevent rust and friction. The pipe is held in place by bolts from stuffing box to lugs or flange on the pipe to prevent their being blown out upon admission of steam. The nuts of these bolts should allow enough play to the working parts.

IV.

THE ENGINE AND ITS PARTS.

THE piston of an engine receives its reciprocal motion by alternately admitting steam to the top and bottom, and permitting escape to the steam which has done its work on the opposite end of the piston to that which is receiving steam. Fig. 36 is a section of a cylinder in which the piston has arrived at the end of the upward stroke, steam being cut off by a single ported slide valve.

Steam from the boiler is admitted by the stop valve or throttle to the valve chest as a receiver. This brings about a pressure in the valve chest equal to that in the boiler and presses the valve firmly to its seat. The faces of seat and valve must be accurately planed and lubricated to allow free travel. In Fig. 37, the valve just covers both steam ports and the engine would be at rest.

A valve complying with this condition is said to have no "lap," meaning that it does not overlap the ports at its ends. Such a valve will have to travel twice the width of a port to accomplish its objects. The distance traveled is equal to the "throw" of the eccentric on the crank shaft from which the valve receives its motion. The "throw" is twice the distance between the center of the shaft and the center of the eccentric disc, the latter being really a substitute for a regular crank.

If the valve in Fig. 37 be moved down to the position in Fig. 38, steam will have been admitted to the upper port a and the piston P will have been forced downward. At the same time, the steam left below the piston from the previous stroke will escape through port b

into the exhaust passage c connecting with the condenser, or the smoke stack, if the engine be non-condensing. In Fig. 38, before the piston has reached the end of the down stroke, the valve will

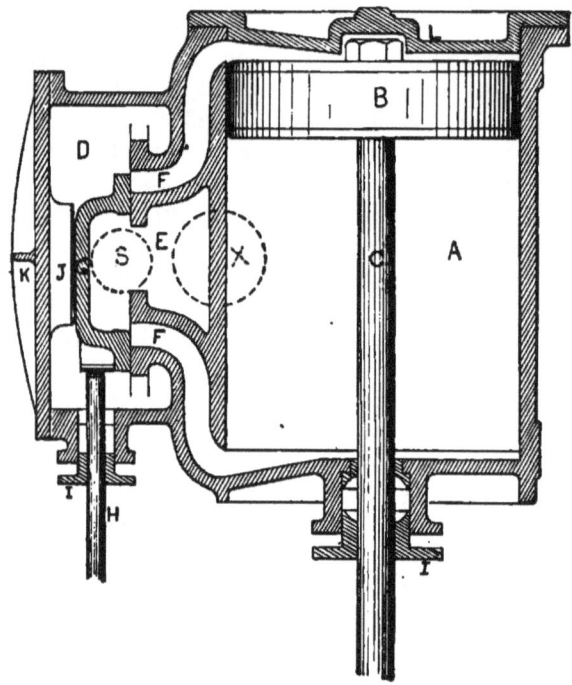

FIG. 36.—SECTION OF CYLINDER AND VALVE CHEST.

A. Cylinder.
B. Piston.
C. Piston rod.
D. Valve chest.
E. Exhaust port.
F. Steam ports.
G. Slide valve.
H. Valve stem.
I. Glands.
J. Guide to relieve valve of pressure.
K. Ribs on steam chest cover.
S. Position of steam pipes.
X. Position of exhaust pipe.

have changed its motion and proceed upward, opening the port b to the steam and connecting the upper port a with the exhaust passage c, as in Fig. 39.

The Engine and Its Parts. 71

A consideration of these diagrams will show that steam must "follow full stroke," and that no work by expansion can be done. For, if we were to set the eccentric in such a way as to cut off the steam from the upper port before the piston has reached the end of the down stroke, the exhaust port would likewise be closed and the lower end of the cylinder left full of steam. The piston would continue to descend until the back pressure on the under side equaled the pressure of the expanded steam on the upper side of the piston, the

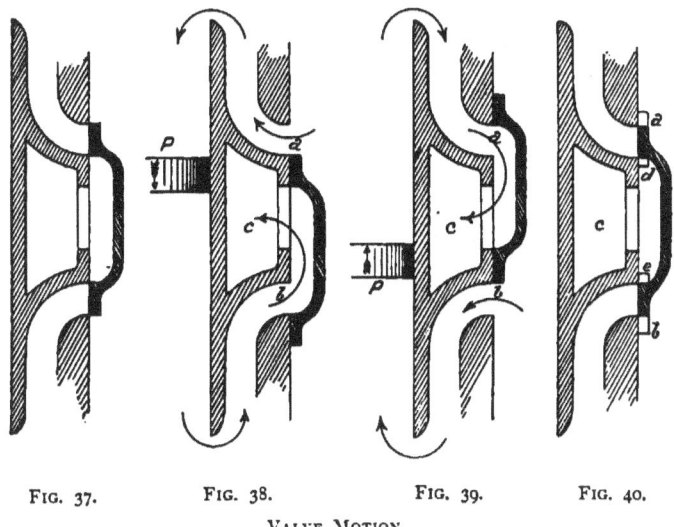

FIG. 37.   FIG. 38.   FIG. 39.   FIG. 40.
VALVE MOTION.

engine coming to a standstill. To work expansively, this simplest form of the slide valve must be modified by the addition of "lap," that is the extension of the valve at its external extremities, so that the steam port may be closed before the exhaust. But this carries with it an error in the opposite respect.

In Fig. 40, the lap added is shown by the extension a and b. Now while a will close the upper steam port before the piston has completely descended, the corresponding lap b at lower end of valve will fail to open the lower steam port in time for the return stroke.

We must therefore "set the eccentric ahead." That is, the valve should be given such motion relative to the piston, that it will have traveled far enough on its upward stroke to open port *b* when the piston is ready to ascend. The stroke of the valve relative to that of the piston is under our control, for the eccentric disc on the engine shaft, from which the valve receives motion, can be shifted around the shaft, or "set ahead" as we desire. This will alter the stroke of the valve in comparison with that of the piston. The arc through which the eccentric is set ahead to meet the addition of lap is called the "angular advance." By putting the valve ahead to adjust the opening of the steam ports, we will introduce a new error on the interior of the valve. From Fig. 40, it can be seen that if the

 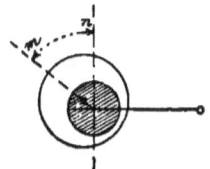

FIG. 41.—ECCENTRIC DISC.

A. Center of engine shaft.
B. Center of eccentric when up.
C. Center of eccentric when down.
B. C. Throw of eccentric.

m. n. Angular advance of eccentric beyond right angle with crank to meet the "lap" of a valve.

valve be set ahead enough to open steam port *b*, the exhaust on top will open too soon by that amount before the piston has descended, and no work could be done by expansion on the upper side of the piston, as the steam would escape through the exhaust. To keep the latter port closed until the end of the downward stroke, we must add lap *d* on the inside of the valve, and the same amount at *e* for the return stroke. If the same amount of lap be added on the exhaust side of the valve as on the steam side, it would close the exhaust at the time steam is cut off, which would be too soon and cause too great back pressure or "cushioning." For this reason the lap on interior of valve must be less than that added to exterior. The eccentric is also set ahead a little more than necessary to open the steam port a little before the extreme end of stroke, so that the piston may be brought gradually to a point of rest before altering its motion, by the cushioning of the steam. The valve is then said to have "lead." Where the clearance spaces are large, closing the exhaust in advance will supply the required cushioning, which should be equal to the terminal pressure upon the piston. In such cases "lead" is not necessary.

To REVERSE an engine, it is necessary to arrest the valve and change the direction of its travel, thereby reversing the admission and escape of the steam on the head of the piston. For example, the steam is entering on top and the exhaust escaping through the lower port. If we suddenly shift the valve, so as to close the steam port above and place that part of the cylinder in connection with the exhaust and at the same time close the exhaust below and open the steam port instead, the piston will come to a stop and will ascend instead of completing the downward stroke. This will cause the crank to turn in the opposite way and the screw shaft with it.

To SHIFT THE VALVE and reverse its stroke, a second eccentric is required. By throwing the ahead-eccentric out of action and the backing-eccentric into action, the result is accomplished. This is done by the intervention of the eccentric gear. In Fig. 42, the various parts of a small launch engine are explained by the lettering attached. The engine is going ahead, the ahead-eccentric giving motion to the valve through the rod E, link block B in the link L, and the valve stem V. To reverse, the handle H of the lever is moved to the left to H^1. This will draw the link L to the left in the dotted position, and bring ahead-eccentric E to E^1 and the back-eccentric F into the position of E in the illustration, at the same time drawing down the block B and shifting the valve to suit. This block B is attached to the valve stem by a pin, but the link slides back and forth on the block. Now, when rod F comes into line with the valve stem V, the latter will be actuated by the back-eccentric. This is keyed to the shaft in a position so related to that of the ahead-eccentric, that upon being thrown into gear the valve will travel in opposite direction. If the link be thrown over half way, or

FIG. 42.—REVERSING LAUNCH ENGINE.

A. Cylinder of inverted direct acting launch engine.
K. Cast iron standard or frame.
T. Bed plate screwed down to keelsons.
U. Engine keelsons in bottom of boat.
C. Valve chest, steam pipe not seen.
X. Exhaust from valve chest.
Z. Guides for piston crosshead.
N. Connecting rod from crosshead to crank.
V. Valve stem, connected by pin with block B.
B. Link block, having free travel in link.
L. Link for throwing eccentrics in or out of gear.
R. Connecting arm between link and reversing lever.
H. Reversing lever with notched quadrant and latch.
E. Ahead-eccentric rod in action.
F. Back-eccentric rod out of action.
S. Eccentric straps in which the eccentric revolves.
M. Eccentric disc keyed to crank shaft.
G. Clutch coupling to line or screw shaft.
P. Feed pump worked by eccentric on shaft.

[74]

in "mid-gear," the valve will cover both steam ports, and the engine comes to a stop.

If thrown over between mid-gear and full-gear ahead or astern, the travel of the valve will be shortened proportionally, opening and closing the ports sooner, admitting less steam than when in full-gear. The effect is the same as diminishing the throw of the eccentrics. Of course the saving in steam would be accompanied with a loss in power, as the mean pressure would decrease. The engine would therefore be going at reduced speed.

In practice the valve is so adjusted as to admit steam from half to two-thirds the stroke, which is all that can be done under the circumstances. Where greater expansion is required, it is necessary to resort to a third eccentric, giving motion to a separate or "expansion valve" seated on back of the slide valve, cutting off the steam from ports in the slide valve at any desired fraction of the stroke. When the fraction can be altered at option, the expansion valve is known as a "variable cut-off." Such alteration is often effected by right and left screw threads on the expansion valve stem, enabling two separate valves to be drawn closer or spread apart by turning the screws, in that way changing the travel of the expansion valve above the slide valve.

For high pressures, "piston valves" such as shown in the Wells Balance Engine, take the place of the slide valve, as they are "balanced" by the steam in the chest surrounding them, thereby preventing cutting the valve seat.

THE INDICATOR is an invaluable instrument, by means of which we obtain a graphic delineation of the internal working of the cylinder. The first apparatus of the kind was invented by James Watt. From the "diagram" taken while the engine is at work we are able to deduce: The initial pressure on the piston; the pressure at all portions of the stroke, during which we follow up with steam; the exact point of cut-off; the reduction of pressure due to expansion during remainder of stroke terminal pressure; counter-pressure in non condensing engines; vacuum of condensing engines; cushioning; lead and mean effective pressure for entire stroke. In connection with the mean pressure, stroke and number of revolutions, we are enabled to ascertain the power developed. The mean power, compared with the steam supplied from the boiler, gives us the cost of power in steam, and if referred to the coal consumption, of power in fuel. The diagram enables us also to determine whether the valves are set and work properly; whether steam and exhaust passages are of right size; whether there is any leakage; the loss between boiler and engine pressure; efficiency of jacketing, and of expansion in one or more cylinders; to apportion the work done in the cylinders of compound engines, etc. The in-

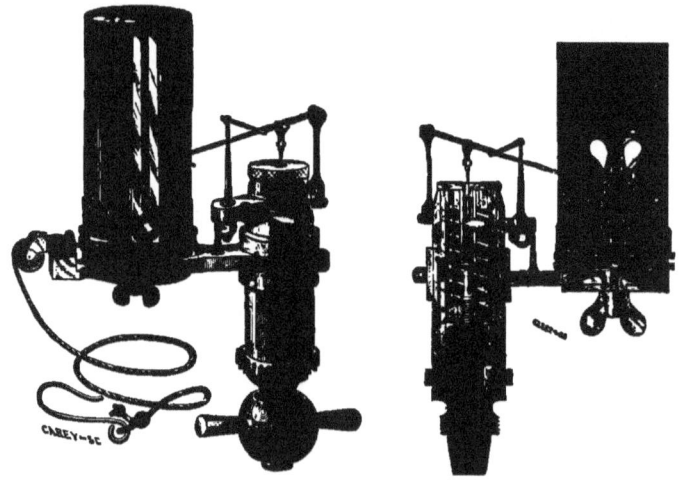

FIG. 43.   FIG. 44.

THOMPSON IMPROVED INDICATOR.

dicator is, in fact, the stethoscope of the engineer. Without it we would be working in the dark, and our stock of information would be limited and untrustworthy. No intelligent reading of the economics of the engine would be within our control.

The operation of the indicator can be described with the aid of Figs. 43 and 44, which are outside and inside illustrations of the Thompson Improved Indicator, manufactured by the American Steam Gauge Co., of Boston. It consists essentially of a small cylinder and large drum rotating on a spindle in a connecting arm. Steam from the engine is admitted into the lower end of the small cylinder, and presses upon a piston with a resisting spring above it. As the piston is forced upward, a rod is driven out of the top of the cylinder. To the head of this rod is attached a lever containing a pencil at its extremity. The pencil point presses against the large or "card cylinder" of the instrument. This is caused to rotate back and forth, a part of a revolution in one direction, by means of a cord attached to a suitable part of the engine, and the return in opposite direction by means of a spring in the base of the card cylinder. A piece of paper is wrapped around the card drum and held in place by vertical fingers screwed to the base.

If steam is admitted under the piston of the instrument, the pencil will ascend, marking a vertical line on the paper cylinder, if the latter is not caused to rotate. If the pencil remains fixed, while the paper cylinder is rotated, the line marked by the pencil will be a horizontal. If the pencil ascends or descends, and the paper cylinder is rotated at the same time, the line drawn by the pencil point will be a curve, the shape of which depends upon the relative amount of vertical motion of pencil and rotative motion of the paper cylinder.

Supposing the instrument attached and in working order, the consideration of an actual diagram will further a clearer understanding. Fig. 45 is the sheet of paper removed from the paper cylinder and flattened out. The o line, or "atmospheric line," is obtained by causing the cylinder to rotate before the admission of steam, the atmosphere pressing alike on top and bottom of the piston. If during the operation the pencil is forced above that line, it will show a pressure greater than that of the atmosphere. If the pencil descends below the o line, it is due to a pressure less than that of the atmosphere on the piston of the instrument. Steam being admitted into the engine cylinder and at the same time from it into the instrument, the piston of the latter will commence to rise, describing on the paper cylinder a vertical line from a up to b. At b, the paper cylinder, having rotary motion coincident with the travel of the engine piston, starts to revolve, and the pencil describes the line b to c. At c, the valve in the valve chest of the engine closes, the rest of the stroke being effected by expansion. The pressure in the cylinder gradually falls and the pencil descends while the paper cylinder continues to rotate. These two motions combined produce the curve *c d*. At *d*, the exhaust port opens and the pencil quickly drops to *e* in consequence of the withdrawal of pressure. The engine piston now starts on the return stroke and the spring in the paper cylinder causes the latter to reverse its rotation, the pencil describing the line *b a* back to the beginning of the diagram.

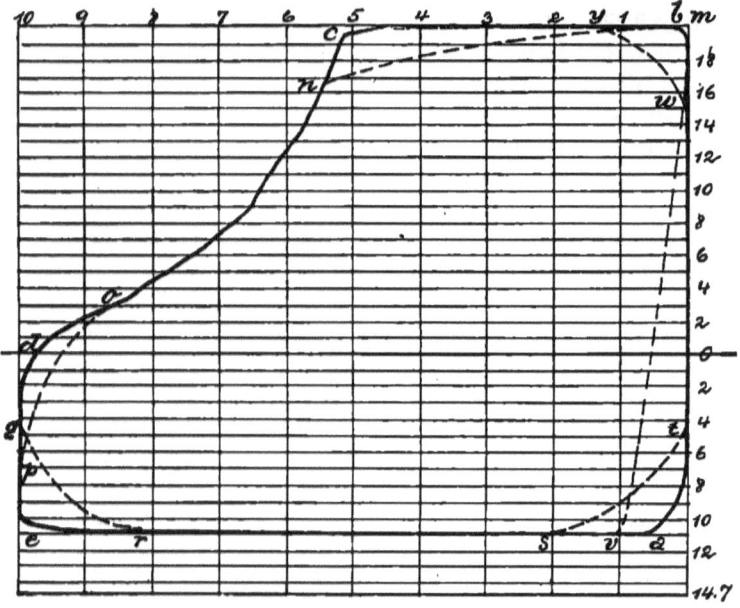

FIG. 45.—INDICATOR DIAGRAM.

The line from *a* to *b* is manifestly the "receiving line," representing the rise in pressure from the partial vacuum of the condenser or exhaust side of the piston up to the time the piston begins to move in answer to the admission of live steam. From *b* to *c* is known as the "steam line." Its length represents the time of following full steam. It does not rise or fall because the pressure remains constant in the engine cylinder and therefore also under the piston of the instrument. From c to d, the pencil drops as the pressure in the cylinder decreases with expansion. It is called the expansion line." Nearly at the end of the stroke, upon opening the exhaust port, the pressure suddenly drops from the terminal pressure on the piston at d to that in the condenser at e. This is called the "exhaust line."

The return stroke of the engine piston moving against the pressure in the condenser, is shown from e to a and is horizontal, because the pressure on the exhaust side of the engine piston, with which side the instrument is in communication, is constant. This is called the "back pressure line," or incorrectly the "vacuum line." In non-condensing engines, it will be coincident with, or but little above the atmospheric line as they exhaust into the smokestack against the pressure of the atmosphere and any back pressure from the escape pipe.

The numbers on right side of diagram are a scale of pressures in lbs. The numbers along the top represent the length of the stroke in feet. It will be seen that the initial piston pressure was 20 lbs. and the vacuum on exhaust side of piston 11, equal to a total unbalanced pressure of 31 lbs. per sq. in. of piston area at beginning of stroke. The cut-off is a little more than half the stroke. The rounding at d is the "lead" on the exhaust; the exhaust port having opened a little before the piston arrived at the end of its stroke. The round at a represents the amount of cushioning due to closing of exhaust port before end of stroke. Without lead the corners would have been angular. The round at e at commencement of return stroke is the back pressure still in cylinder. The scale of pounds pressure will depend upon the strength of the resisting spring, which is made to allow the piston of the instrument a certain travel for every pound of pressure put upon it. The scale of the stroke depends upon the extent to which the paper cylinder revolves per stroke of engine piston. The "mean pressure" for the entire stroke is found by measuring across the diagram at intervals and dividing by the number of intervals.

THE HORSE POWER is obtained by multiplying the mean unbalanced pressure per square inch upon the piston by the area of the piston in square inches and again by the speed of the piston in feet per minute and dividing this product by 33,000. If A = area of piston in inches, p = mean pressure, v = velocity of piston, then the

$$I. H. P. = \frac{A p v}{33,000}$$

The divisor is supposed to be the weight which a stout draft horse can raise one foot high in one minute of time, hence the term horse power as applied to the work of an engine. The speed of the piston per minute is equal to the stroke in feet multiplied by the number of strokes per minute.

A PERFECT DIAGRAM should follow Marriotte's law previously explained, which serves as the standard of comparison. It can be drawn in according to the directions given for apportioning the lengths of the ordinates of the hyperbolic curve in inverse ratio to the volume occupied by the steam.

Had the diagram Fig. 45 shown the corner cut off like the dotted line *s t*, it would mean that the exhaust closed too soon, occasioning excessive cushioning and lowering the mean unbalanced pressure. Had the upper corner been cut off, as at *w y*, the steam valve would open too late. Had the steam line dropped to *m n* instead of being nearly horizontal, it would indicate cramped passages and port or the throttle not opened wide enough, for the pressure in cylinder would not have been maintained as it ought to have been while following up. Had the exhaust corner *o p* been cut off, the exhaust port would have opened too soon. Had the corner *g r* been cut off, the exhaust port would have opened too late. If the lead to the steam side were excessive, it would show in a line like *v m*.

The Thompson Improved Indicator is adapted to extreme high pressures. The usual piston has an area of ½ sq. in., which with a 100 lb. spring, provides for indicating pressures up to 250 lbs. By substituting a smaller piston the capacity of the spring can be increased to 500 lbs. Much depends upon the free working of the instrument. It stands to reason that if the moving parts are hindered in their play, or if the springs are not true in their action, the results will be deceptive.

Before applying the indicator, it should be taken apart and carefully cleaned and oiled. Each piece should be proved separately.

Put them together without the piston spring and lift the pencil to see if it will fall clear. Then put in the spring and connect. Give it steam to warm and expand all parts. Lead should not be used in making tight connection, as it is liable to get into the casings. A suitable coupling comes with the instrument. The lighter the spring used the higher will the diagram be and more accurate measurements can be taken, as the scale of pressure will be larger. The diagram should have a height of say two inches. The following rule will give the maximum pressure to which each spring should be subjected. Multiply the number or scale of spring by $2\frac{1}{2}$, and deduct 15 for the vacuum allowance. Example: 40 lbs. spring $\times 2\frac{1}{2} = 100 - 15 = 85$ lbs., the maximum pressure for the 40 lbs. spring. For smaller pressures the spring can of course be used.

The guiding pulley over which the cord of the paper cylinder is led can be adjusted to any direction by a set-screw as shown in the illustrations. The lead pencil should be hard to insure sharp lines. Only fine watch oil must be used to prevent gumming. After using take the indicator apart and clean carefully.

When no provision has been made for attaching the indicator, holes must be drilled and tapped in the side of the cylinder near each end, so that when the piston is at the end of its strokes, the holes shall be about midway in the clearance space left.

If obstructed by the piston, steam will be cut off from the instrument. The tap should be for half-inch pipe. Chips from drilling should be blown out of the cylinder by turning on steam. If clearance space is too small, drill directly into the head. Cocks should obstruct the passage of steam as little as possible. By using a "three way" cock leading to both ends of the cylinder, one indicator will serve for taking diagrams from both ends by admitting steam accordingly. This will do away with changing the cord connection rotating the paper cylinder. The motion of this cylinder could be got by attaching the cord wound around the base to the crosshead of the engine piston. Manifestly this would afford too great motion and would cause the paper cylinder to revolve several times. The stroke of the engine must be reduced by the intervention of some combination of reducing levers.

The Engine and Its Parts.

Numerous devices exist for this reduction, the chief point being that the motion of the piston crosshead should not be in the least distorted by the intervening mechanism, as is too frequently the case. The most correct and convenient device which transmits the true motion of the crosshead to the paper cylinder is the Bacon Pantograph, of which we give illustrations in Figs. 46 and 47.

The following description of method of attaching the indicator cord by "Chordal" appeared in the *American Machinist:*

"It is manufactured by the American Steam Gauge Company of Boston. It consists of a lazy-tongs system of levers. The long

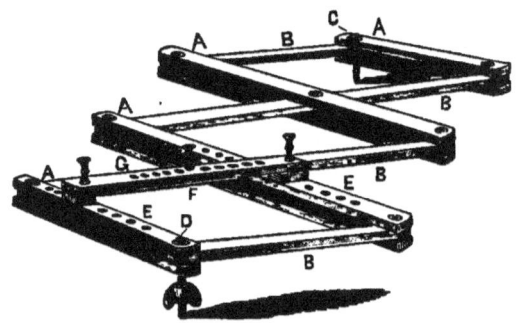

FIG. 46.—THE BACON PANTOGRAPH.

levers are of cherry wood, 16 in. between centers, 1⅛ by ₁⁄₁₆; those marked B being single strips, and those marked A being double strips. This makes the thing very stiff and substantial. The pivots should be got up in good style, and the pivot holes bushed. The hitch strip G should be arranged so that it may be shifted in the holes E, and bring a hitch pole, F, in a line passing through pivots C, D. The end pivots C and D should have a projection below of, say, 2 in., with the end somewhat pointed. Any one who attempts to make one of these things will have fun. The least variation in the location of the pivot holes will cause the levers to refuse to act. No dimensions are essential; and if the thing will close up nicely, and open out nicely, it is all right; if it won't do both, it is all wrong. The engine crosshead must have a vertical hole in it somewhere, so that pivot C

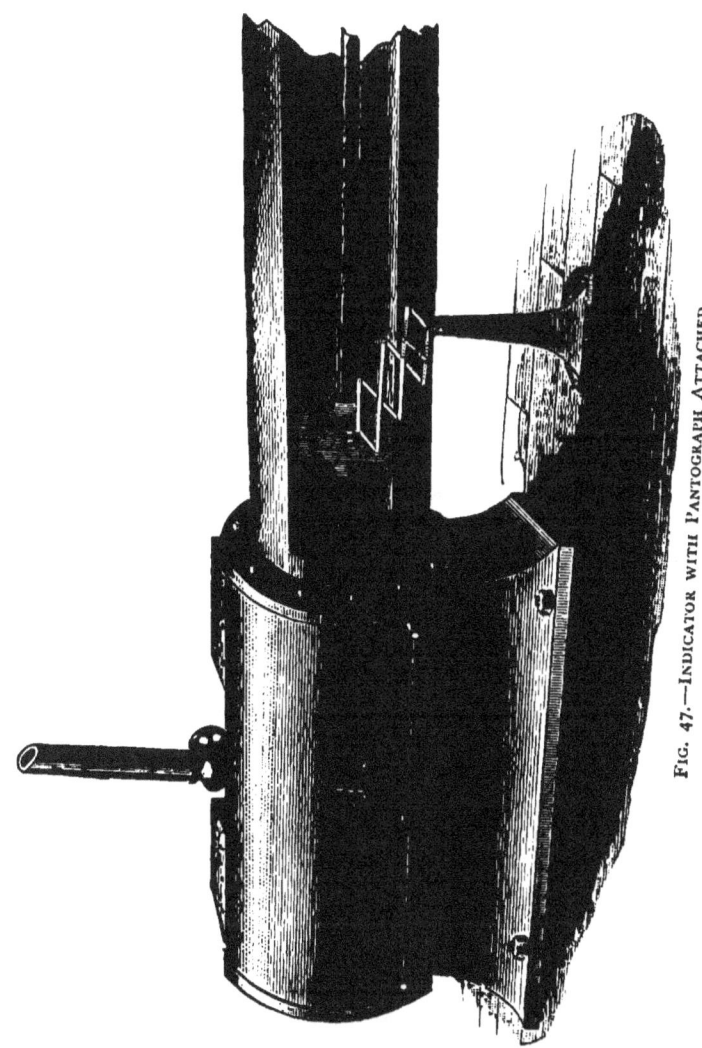

Fig. 47.—Indicator with Pantograph Attached.

can be dropped into it. A stake must be set in the floor near the guides, having a socket for the pivot D in its top. The stake socket must be level with the crosshead socket, and must be directly opposite the crosshead socket when the latter is at mid-stroke. The indicator cord is hooked to the center peg F, and the stake should set at such a distance from the guides that the cord will lead off parallel with the guides. Otherwise a guide pulley will be called for. When this ring is in motion, every point on a line cutting C D has a true motion parallel with the guides, varying in distance from nothing at D to length of a stroke at C. It is only necessary to hitch the cord at a point on this line which will give the right amount of motion to the cord. This point will be near D, and within the range of adjustment of the strip G. This is as neat a device as could be wished for. I have seen men hook on to an engine running at a good gait, without stopping. For a permanent rig on a nice engine, the stake can normally support a neat table top for oil cans and waste."

CONDENSERS are supplied to all large yacht engines, but steam launches are still generally worked against the pressure of the atmosphere. In small boilers it is easier to obtain the requisite strength to withstand high pressure, and it is generally deemed preferable to carry from 10 to 12 lbs. more pressure in the boiler than to go to the expense and weight of attaching a condenser with its pumps and piping. Fresh water for feed can be obtained in a launch at short intervals, and by occasional blowing off, a small supply of fresh water will keep the density low enough for short runs, even if the boiler be filled from the sea. For long runs and sea cruising, however, a condenser is a necessary part of the marine engine. For the development of a given power the boiler pressure can be lower by just the amount removed from the exhaust side of the piston by working against the partial vacuum produced by condensing the exhaust steam in a special chamber for that purpose.

The steam, having done its work in the cylinder, is allowed free escape through a pipe of large sectional area leading to the condenser. There it is brought in contact with cold water pumped in from the sea by the "circulating pump," and precipitated or condensed into water, forming a partial vacuum, as water will occupy only a sixteen

The Engine and Its Parts.

hundred and forty-second part of the volume filled by an equal weight of steam at atmospheric pressure.

The earlier condensers were of the simplest kind, a jet of cold water being forced into a cast iron chest into which the exhaust was received. The jet was taken from the sea, hence the condensed steam mingled with it, and being pumped back into the boiler, supplied water almost as salt as that of the sea itself. In vessels navigating fresh rivers or lakes, the jet accomplished all that was desired, and this form is still commonly used in steamboats which fill up with fresh water at stated periods at the dock. But for sea-going purposes or irregular cruising, where the boiler water cannot be changed as required, a superior arrangement, known as the "surface condenser," has displaced the jet altogether. It was not until 1862 that this new form received the attention it deserved. Numerous patterns have since been evolved, differing in mechanical details, but all operating on the same plan.

At first the exhaust was passed into a series of copper or brass tubes in the casing of the condenser, the cooling water being allowed to drip down through a sprinkling plate above. The steam was condensed in the tubes and ran into a collecting well, from which it was returned to the boiler by means of the "air pump," so called from the fact that it draws both water and air from the condenser. At present, however, the method is reversed, the cooling sea water being forced through the tubes by a "circulating pump," and the steam condensing in the spaces between the tubes by coming in contact with their cool surfaces. The condensed steam drops to the bottom, whence it is withdrawn by the air pump and forced either directly to the boiler as feed, or collected in an intermediate receiver called the hot well, where it serves as suction for the feed pump.

The Lighthall, though extensively used in the merchant service and English yachts, is now being superseded by the Wheeler Improved Surface Condenser with independent air and circulating pumps, capable of being worked and regulated without regard to the main engine. The objections to the ordinary surface condenser are the unequal expansion and contraction of tubes, breaking or rupture of tube heads, leaking of packings at head and the "crawling" of tubes partially or entirely out of their heads, unequal distribution of steam over cooling surface, so that parts of condenser are hot while other parts are cold, insufficient vacuum, condensed steam (feed water) not as hot as it should be, and the liability of the circulating water to leak into the steam space and mix with the water of condensation—thus defeating the objects for which surface condensers are intended. The Wheeler Surface Condenser has none of the above objections, and combines the necessary theoretical qualifications with sound practical features. The tubes are so arranged that they are free to expand and contract without the use of packings of paper, wood or similar materials; there are no ferrules, followers, washers or packings of any kind employed. Plain screw joints are used—the simplest, most durable and efficient tube fastening possible, and always tight. The tubes are straight, of seamless brass tubing, tinned inside and outside. They can be easily taken out and thoroughly cleaned, as their form and the means of fastening them permit of this being readily done. The tube heads do not have to be removed from the condenser for the cleaning or repairing of the tubes. The pressure (and likewise the temperature) of the exhaust steam as it enters the condenser is reduced to a minimum, and is then uniformly distributed over the cooling surface. This, together with a perfect circulation of water in the tubes, produces a more uniform temperature in the condenser, making one portion as efficient as another, and economizing the amount of cooling surface and circulating water. The water of condensation passes from the condenser at the hottest temperature possible. The circulation is active and thorough, consequently a smaller amount of circulating water is required; this feature gives a saving in the capacity and power necessary to work the pump.

Referring to the accompanying sectional illustration, the operation of the condenser is as follows :

The exhaust steam from the engine entering the condenser by the nozzle A, comes first in contact with the perforated scattering plate O, which protects the central portion of the upper tubes from the deteriorating effect of the direct impingment of the steam.

The steam expanding in the spacious top of the condenser,

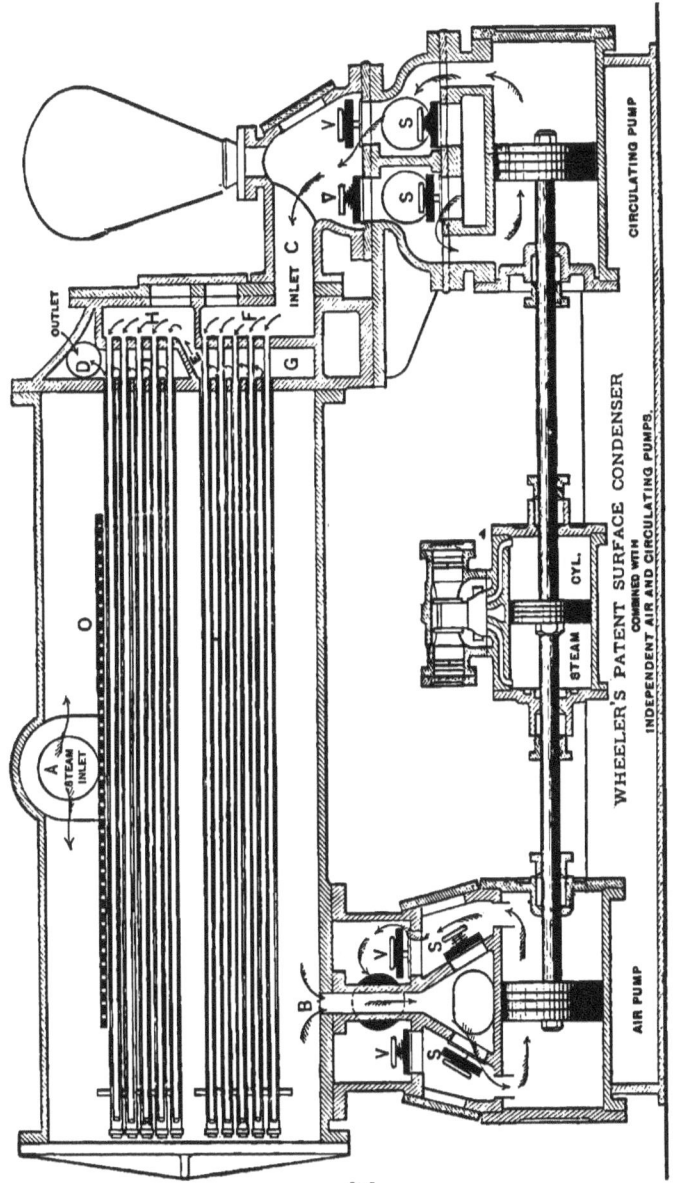

Fig. 48.

WHEELER'S PATENT SURFACE CONDENSER
COMBINED WITH
INDEPENDENT AIR AND CIRCULATING PUMPS

[87]

reduces its pressure and temperature before it comes in contact with the cold tubes. The steam as soon as condensed, gravitates to the bottom, and passes out by the nozzle B to the air pump.

It will be noticed that there is ample room in the bottom of condenser for the water of condensation, so that it cannot come in contact with the cold tubes and become chilled; the hot water therefore passes out at the highest possible temperature—according to the vacuum carried.

The circulation of the condensing or cooling water is as follows: It is pumped into the compartment F through the nozzle C, and

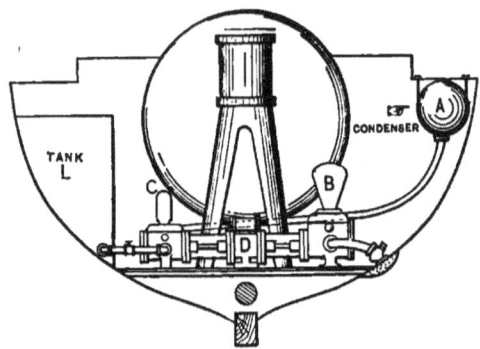

FIG. 49.—SECTION THROUGH LAUNCH.

A. After end of condenser.
B. Circulating pump.
C. Feed pump.
D. Combination pump steam cylinder.

enters the small tubes as shown by the arrows. After traversing the small tubes, the water returns through the annular spaces between the small and large tubes of the upper section in the same manner, and finally passes out of condenser by the discharge nozzle D.

The small tube is expanded into a screw-head which latter screws into the head of the casting at H. This small tube ends within a few inches of the cap at the loose end of the large tube, thereby giving ample space for the water to reverse its direction before flowing back through the annular space between the two tubes. The end of the large tube screws into the casting head D G, so that coarse deep threads and a screw-driver slot can be cut, which admits a tool for screwing up or unscrewing tubes from the tube heads. When necessary to remove the tubes for cleaning or repairs, both small and large tubes can be drawn out from the same end of the

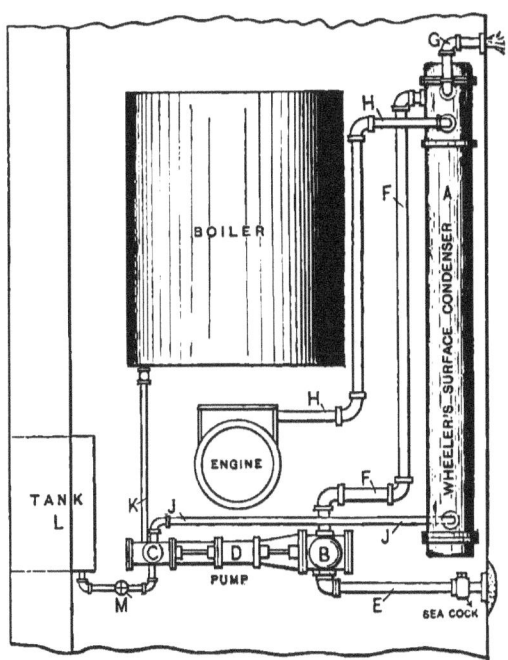

FIG. 50.—PLAN OF CONDENSER AND CONNECTIONS.

A. Condenser.
B. Circulating pump.
C. Feed pump.
D. Combination pump steam cylinder.
E. Suction to circulating pump.
F. Delivery from circulating pump to condenser.
G. Outboard delivery of circulation water.
H. Exhaust steam from cylinder to condenser.
J. Suction from condenser to feed pump.
K. Feed to boiler from feed pump.
M. Suction from tank to feed pump.

condenser. After removing the small tube the large tube is unscrewed and drawn through the hole left vacant by the screw head of the small tube—this hole being a little larger than the thick end of the large tube.

The illustrations given are taken from condensers of very large capacity. For steam yachts, the same principles and arrangements are followed, but the weight and size are much reduced. Thus, for a 40 H. P. engine, the body of the condenser is a 12 in. pipe 7 ft. long. The Wheeler condenser with pumps complete will weigh less than the usual marine condenser without pumps and piping, and in launches can be set up under the thwarts out of the way.

The advantages arising from the use of surface condensation are the increased power derived from working against a partial vacuum on the exhaust side of the piston instead of against the atmosphere, the furnishing of distilled water to the boilers preventing loss of heat by frequent blowing off and scaling at high pressures.

In many small American yachts an "outboard condenser" has been adopted. This consists simply of a pipe running fore and aft along the garboards on the outside of the vessel. In point of simplicity, cheapness and saving of weight this plan is far ahead of the old style "Lighthall" condenser with its heavy cast-iron casing and rigid tubes, and is being extensively adopted. Being in contact with the sea, its whole surface is subject to the same cooling temperature, whereas in the ordinary condenser greater surface must be provided to allow for the heating of the injection water as it proceeds through the tubes. The circulating pump is also abolished, as also the "injection" and "outboard delivery" valves and piping attendant upon the use of the pump. The Hancock "ejector" also takes the place of the air-pump at a great saving in weight, and the same instrument will deliver the condensed steam to the hot well or tank from which the boiler is supplied.

In large yachts a series of outboard pipes oblige the exhaust to pass fore and aft two or three times, insuring large cooling surface in small pipes. They have a block of wood tapered away to a point at their forward end. No power is lost working a circulating pump and no deleterious acids from the decomposition of lubricants and the copper or brass tubes or the "packing" of the latter are returned to the boiler, as is the case with the regular surface condenser.

The "outboard" condenser on anything larger than launch or river yacht, is always in the nature of a makeshift and an obstruction

The Engine and Its Parts.   91

to the free closing of the water in the vessel's run and its access to the screw. The liability to accident is always present and the expense of docking for repairs may more than counterbalance saving in first cost. It is not to be counselled for cruising yachts which will frequently find themselves beyond the reach of docking facilities. For reliable service the Wheeler condenser is to be preferred. In some recent torpedo boats the circulating water is forced by the speed of the vessel, the orifice of a pipe projecting forward through the boat's skin, the travel of the vessel driving the sea water into the pipe. This is supplemented by a steam jet when the speed is insufficient.

In all yacht engines, the surface condenser has hitherto been made to form part of the engine framing, as will be seen from illustrations in these pages. This is still a feature of all English marine engines. But the latest American improvements, such as the Wheeler condenser, are independent attachments, which is preferable. For lake and fresh water river service, the jet will take the place of the surface condenser, as no objection exists to mingling the injection water with the condensed steam.

The use of the condenser enables the engine to work against a vacuum, and is therefore desirable, even if the air pump sends the contents overboard instead of to a hot-well. By using the contents over again, however, as feed for the boiler, there will be a saving of heat as the condensed steam will impart its heat to the injection and leave it warmer than cold feed pumped in from the river. If the latter be muddy, the feed from the hot-well will be partly purified through the distilling operation of the condenser, and less dirt will be fed to the boiler. The jet is lighter, simpler and cheaper than the surface condenser necessary in navigating salt seas.

A compact and excellent arrangement is shown in the illustrations of the Worthington Independent Jet Condenser. The injection cannot be drawn over into the steam cylinder, and the danger of flooding is thus avoided without the use of floats, check-valves or automatic contrivances.

Should the pump from any cause be stopped while the main engine is at work, the vacuum would be immediately destroyed, and

Fig. 51.—The Worthington Independent Condenser.

the injection water stop flowing, as the latter is not forced by a special circulating pump, but is drawn in by the suction of the air pump. No "head" is required to the supply for the same reason, as the air pump will lift water from any point within the limits of suction.

In the illustration, the exhaust steam from cylinder passes in at the elbow on top. The injection enters through the valve above the elbow, and passes over a cone in the pipe, provided with wings which separate the water into a shower of spray to insure complete admixture with the steam. The cone is adjustable by the handwheel shown in the cut, and all strainers are done away with, so that liability to choke up is removed. By simply lowering the cone, any obstructing article introduced with the injection can be freed. The injection water, after it has done its duty in the condenser, can be forced, if required, to a tank or hot-well at an elevation, if it is not to be discharged through a sea valve. This condenser has been fitted with great success to many yachts, notably to the steamer built for Mr. David Bell of Buffalo, whose engine is of the compound type, with cylinders 9 and 12 in. by 12 in. stroke. The exhaust from the steam cylinders of the pump is turned into the condenser by a pipe, so that the air pump will also work against a vacuum.

No. of Condenser.	Diameter of Steam Cylinders.	Diameter of Water Cylinders.	Length of Stroke.	Diameter of Engine Exhaust Pipe.	Diameter of Injection Pipe.	Diameter of Delivery Pipe.
1	3	4	6	4	2½	2
2	4	5¾	6	5	3	3
3	5	7	10	6	4	4
4	5	7	10	7	4	4
5	6	8½	10	8	5	5
6	6	8½	10	9	5	5

A feed water heater can be introduced between engine cylinder and the condenser, so that the heat from the exhaust steam can be utilized in raising the temperature of fresh feed, which is forced through the piping of the heater by the feed pump.

The state of the vacuum is indicated by a mercury vacuum gauge constructed as explained for steam pressure, but reversed by connecting the basin with the condenser and commencing from the bottom with the scale. According to the presence or absence of pressure in the condenser, the mercury in the basin will ascend the tube until balance is established. A rise of two inches in mercury is equal to one pound per square inch, and would indicate one pound more

FIG. 52.

pressure in the condenser than a perfect vacuum. A rise to 28 on the scale would indicate a difference of about 14 lbs. between the pressure inside and outside the condenser. The vacuum would be —14, that is, 14 lbs. below atmosphere. It is usual to designate the vacuum by the scale of inches; in this case it would be called 28. The glass need not be 30 in. long, as the engine is supposed to maintain at least 16 to 17 in. vacuum, so that 14 in. will be tube enough to cover the range likely when working the engines. Vacuum gauges are also frequently made on the Bourdon principle. It is customary to have all gauges in the engine room set in one frame, so that ready inspection can be made. Such a frame will include a clock, steam and vacuum gauges and a revolution counter, showing the number of turns made by the shaft. From the latter the "slip" of the screw is ascertained, being the difference between the distance the vessel would have steamed had the screw worked in a solid and the lesser distance actually made. The speed of the engine is also regulated by the counter. It is operated by levers connecting with the engine, and a worm and series of geared wheels inside the case.

V.

THE SCREW.

IT is not necessary to include the paddle-wheel as an instrument of propulsion, for it is obsolete in yacht building practice. A few side-wheel steam yachts are still in commission in British waters, but the superior efficiency of the screw and its adaptability to the lightest draft are so well proven that the paddle-wheel calls for no attention. With the high piston speeds now becoming all but universal, the paddle-wheel is out of question, even if it were not objectionable on other scores, such as great weight, cumbrous boxes overhanging the side, loss due to oblique action of blades and irregular dip in a seaway, etc. In theory the feathering paddle-wheel, in which the blades are retained nearly in a vertical position during the revolution by means of an eccentric on the shaft with arms to the swinging blades, is the most efficient propeller in smooth water. A wheel can be used smaller than when the blades are fixed, as in the common radial wheel, as the oblique action upon first dipping and finally emerging from the water is done away with. The feathering wheel is therefore adopted where higher piston speed is desired, and is more suitable for rough water. But the weight and complication of the feathering gear and its friction are objections which leave the advantage with the screw.

The operation of a screw can be likened to that of a common carpenter's screw progressing into wood in answer to rotary force applied through the screw driver. In a vessel the water takes the place of the wood, being really an endless nut in which the propeller revolves. If we cut a longitudinal piece from the carpenter's screw we have a representation of the common or "true screw" of ship propulsion. It is, however, not necessary in practice to have a section so long that the thread shall wind once round the central spindle or hub. A short part of a complete revolution is enough, as resisting area to the water must be preserved. The efficiency of the screw is further increased by introducing one or more intermediate threads, as if we wound additional spirals around the carpenter's screw. Each blade of the ship screw represents a short piece of as many threads as there are blades. The distance along the axis of a screw required to complete one convolution of the thread is the "pitch." In a three-bladed screw three threads are wound about the axis within the same distance along the axis.

The "axis" is the imaginary fore-and-aft center line through the shaft upon which the threads are wound.

The "radius" of a screw is half the diameter or the distance from center of shaft to periphery of blade.

The "length" of a screw is the fore-and-aft distance taken up by the blade, and varies with the pitch and extent of periphery or distance along the outer edge of the thread.

The edge of the blade which strikes the water first is the "leading edge," the after one being the "following edge."

When the face of the blade has a curvature increasing aft from the leading to the following edge, it is said to have an "expanding pitch longitudinally." That is to say the pitch of the thread has been increased, so that the following part, finding the water already driven astern by the leading part, shall preserve its effect upon the escaping column.

When the pitch of the periphery differs from that of the blade near the hub, the screw has a varying pitch radially. At the hub the blade is moulded more nearly fore-and-aft in line with the vessel's keel, to supply metal enough for strength. If the same pitch were carried up to the periphery, the screw would be too "coarse," and would churn the water without driving it astern. The pitch at the hub is therefore diminished gradually toward the periphery, and the blade made wider where it is most efficient. To do away with the useless churning near the axis, the central part of some screws, as in the Griffiths propeller, is filled in by a large spherical hub, only the effective portion of the blades projecting. The pitch toward the axis is also made coarser to allow for the slower speed near the center. A given pitch and speed may be effective at the periphery, but the same pitch at a slower speed would fail to do its share of propelling.

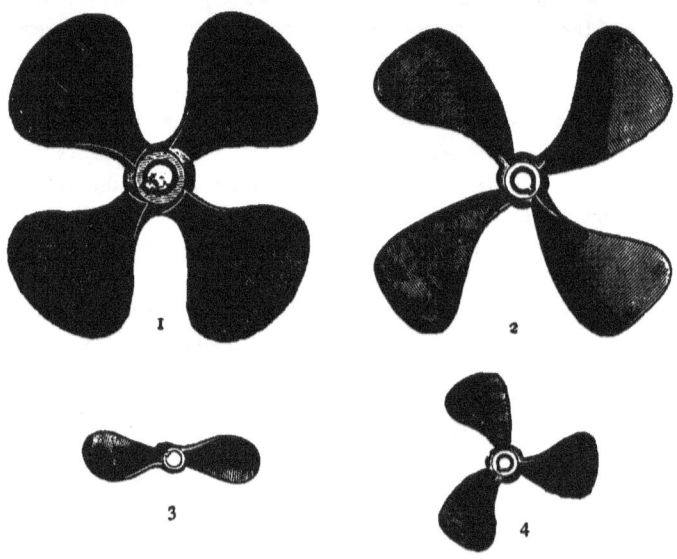

FIG. 53.

No. 1. Towing wheel with large area. No. 2. Four-bladed speedwheel for yachts.
No. 3 and 4. Two and three-blade wheels for launches, according
to practice of Chas. P. Willard & Co., Chicago.

The after face of a screw is called the "driving face," the forward surface being the "dragging face." This face is moulded with a view to supplying the requisite metal to the blade, and its form is governed by that of the driving face. Chief Engineer F. B. Sherwood, U. S. N., reversed the screw while testing a Herreshoff built yacht and obtained equal results from the convex or dragging face as from the driving face, due no doubt to the close similarity of the faces in a small screw with thin blades.

The "oblique area" of a screw is the sum of the actual areas of the blades. The "effective area" is the area of the blades projected upon a thwartship plane. The "disc area" is the area of the circle described by the screw's diameter. The "helix" is the spiral described by a point on the blade during its onward progression. If the screw were working in a solid, this would correspond to the thread.

The "center of pressure" of a screw is located from the axis at a distance equal to the radius of a circle having half the disc area. The pressure outside of a circle drawn with such a radius will equal that within the circle, providing the pitch is constant from hub to periphery, so that the velocity of the columns of water driven aft will be the same all through.

The "*apparent* slip" of a screw is the difference between the speed of the propeller and that of the ship. If the screw worked in a solid instead of a yielding medium, the two speeds would be alike; but as the water slips away from the screw it follows that the screw must make an increased number of revolutions for the production of a given speed in the vessel.

The reacting pressure upon the screw is transferred through the shaft to special bearings made to receive this pressure or "thrust," the bearings being known as the "thrust bearings" in contradistinction to the ordinary bearings which support the shaft in line. The amount of thrust is measured by a dynamometer, an instrument containing levers and weights with scale. The thrust multiplied into the distance the vessel moves in a unit of time, shows the actual power utilized in driving the ship.

If all the power applied to the piston were transmitted to the water through the propeller, the total pressure upon piston and the thrust of the propeller would be equal. The theoretical thrust in pounds would therefore be:

$$\frac{\text{Total unbalanced pressure on piston in lbs.} \times 2 \text{ stroke in ft.} \times \text{No. of Revs. per Min.}}{\text{Pitch of propeller in ft.} \times \text{No. of Revs. per Min.}}$$

The difference between this and the actual thrust as per dynamometer, shows the amount lost in friction of engines, propeller and load, resistance of edges of propeller blade, working air and circulating pumps, etc. The loss from slip is independent of this. Apparent slip is expressed in percentage of speed. If the vessel be moving at 8 knots and the screw at 10, then

$$\frac{10-8}{10} \times 100 = 20 \text{ per cent. slip.}$$

Should the after lines of a vessel be so full as to draw a current in the wake of the vessel in the direction in which she is moving, it might be possible that the vessel would move faster over the ground than can be accounted for by the revolutions of the screw, as the latter is working against a fluid with onward motion and gains proportionately in its thrust. This phenomenon is termed "negative slip," a term which is only one of comparison to express the relation between speed of ship and screw. In truth no screw can have negative slip of its own making. This so-called slip represents a wasteful expenditure in power in producing the following current and is seldom observed. In yachts with clean runs it never occurs.

"*Real* slip," the velocity with which a column of water is thrown astern from the screw, is not to be regarded as an evil characteristic, but on the contrary, indicates the amount of beneficial work the screw is accomplishing. Absence of real slip is a sign of inefficiency. Water being a yielding medium, and action and re-action always alike, it follows that the thrust transmitted to the bearings will vary with the velocity with which water is driven astern by the screw.

If we imagine the pitch of a screw steadily diminished, its work for a given number of revolutions will decrease as also the velocity with which the water is driven astern. Ultimately when the screw has been flattened out into a plain thwartship disc, there will be no slip at all and no thrust, and the vessel would remain stationary in absence of slip. But as the pitch is decreased less power will be required to give it rotation. Starting with equal steam it follows that a refinement of pitch permits a greater number of revolutions, by which the speed of the water driven astern will be accelerated, and in that way make up for the finer pitch.

From this the conclusion would be reached that fine pitch with increased number of revolutions is equally as effective as coarse pitch with a smaller number of revolutions. Within limits this is the case, but the intermediate loss in the working of the engine increases with the number of revolutions, and the limit would make itself evident in practice. Too fine a pitch driven at great speed will prove an error on one side, just as too coarse pitch and too few revolutions on the other. The only safe guide is the comparison of speeds of vessel with a given Indicated Horse Power. The whole problem of the screw in practice is so complex and its efficiency a compromise between so many antagonistic requirements, many of which we cannot measure or even follow up, that nothing but actual test can be depended upon.

In general, coarse pitch, large area and slow revolution is advantageous for power such as required in a tug; but for speed, finer pitch, smaller area and rapid revolution has been the rule. Of late, however, the pitch in high speed vessels has been increased with a gain in efficiency but a greater load upon the engine, as the number of revolutions has to be preserved to secure some of the greater speed due to coarser pitch. The blade area has also been reduced, as it has been shown that the leading portion is the effective part. Too small area will increase the apparent slip and too large area adds to the friction of screw and wastes power. The number of blades does not seem to affect efficiency of the screw, but three or four blades work smoother than two, causing less vibration.

Experiments made by Messrs. Yarrow & Co., on one of their high speed torpedo boats in 1879, warrant certain definite conclusions. The maximum speed reached was 21.9 knots or over 25 land miles.

The resistance at speeds over 18 knots does not increase as rapidly as below that figure. The elasticity of propeller blades greatly increases efficiency, thin blades with sharp anterior edge having increased the speed of the launch from $17\frac{1}{2}$ to 19 knots. Screws showing the least variation in slip at different speeds are the most efficient, and those having small slip at low speeds are the worst. In two cases like results were obtained from two screws, one of large diameter and fine pitch and the other of small diameter and coarse pitch. Propellers best suited for high speeds are not as well fitted for low speeds. Finally that there is a propeller best suited to the conditions and failure to select properly may have important bearing upon speed and even affect the engine performance.

In the steam yacht Celia, a change in screw accomplished the same speed with two-thirds the number of revolutions, representing a great saving in wear and tear to the machinery. Propellers placed under the body of the vessel do not give as good results as if placed abaft the sternpost. By extending them some distance beyond

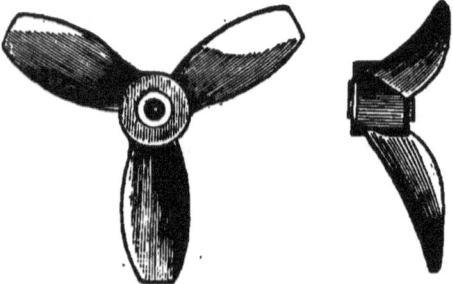

FIG. 54.—THE THORNEYCROFT SCREW.

better results are obtained than when placed nearer to the post, probably because working in water at rest and not interfering with the closing up of the run.

Summing up all that we know concerning the screw, still leaves us without precise information upon which to formulate directions, and nothing but experiment remains to discover the propeller most suitable to any vessel and the conditions under which it is to work. True screws have given as good results as any of the modified forms.

The celebrated Thorneycroft propeller has "dished" blades, that is bent aft. They also curve outward to prevent the dispersion of the water and cause it to be thrown aft in a solid column. The screw is highly advantageous to speed, adding a seventh over the

The Screw.

speed obtained with the Griffiths and other ship screws, the vibration being also reduced.

The Giant propellers of Bliven & Co., New York, are built either two, three, or four blades, ranging in diameter from 15 in. and over, and any pitch desired, and are constructed of the following metals: Phosphor bronze, hydraulic metal, composition valve metal, brass, annealed steel, cast steel, gun metal, car spring metal or cast iron.

The pitch is a true one from the periphery to the hub, and every part of the blade travels the same distance ahead in each revolution, thus making the whole surface effective. The blade has a relief or compound pitch, one-third of the distance from the hub to the end

FIG. 55.—THE "GIANT" PROPELLER.

of the blade, by reason of which the blades near the hub are relieved of the dead water which is forced out behind the screw and compressed into a solid mass. When the screw is in motion, the compressed water behind it always has a tendency to follow the screw, thus compensating for loss by friction and slip.

In a well modeled yacht these screws will save much loss from friction or slip. In several tests made under different conditions in a steam yacht with one of these screws, 6 ft. diameter, 9 ft. pitch, a gain of three per cent. was made over the speed of other screws.

They leave the water behind comparatively smooth, and do not churn it up, neither does a vessel propelled by this style of screw have a tendency to settle aft, and there is less vibration and trembling on the vessel.

The Duncan propeller, made by Messrs. Ross & Duncan, of Glasgow, Scotland, is very favorably known for its excellent results.

The blades are formed to a combination of curves, and are not of the plain helical construction. The chief peculiarity is a quick curvature near the tips, concave to the driving side, as can be seen in the illustration.

This form was adopted, because all propellers act with a certain amount of slip, necessary to give the propulsive reaction, and there is always a corresponding amount of centrifugal action on the water.

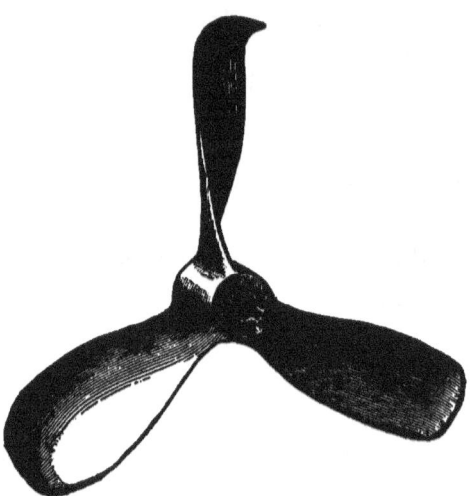

FIG. 56.—DUNCAN'S PATENT SCREW PROPELLER.

The water is drawn in at the center and driven outward along the blades. If the propeller were replaced by a plain paddle the action would be purely centrifugal and no propulsive effect would be given. But all outward motion given to the water is so much loss of power, as it is not driven astern. That propeller which drives the water directly aft is the most efficient. The curvature mentioned in the Duncan screw is introduced with a view to counteracting the centrifugal tendency and to convert the outward motion of the water into an equivalent motion aft.

According to experiments conducted by Chief Engineer Isherwood upon the steamer Lookout, he estimated that 70 per cent. of the work done by the engines was utilized in propelling, 30 per cent. being the intermediate loss between cylinder and screw due to the power requisite to work the engine and its waste.

For many years it has been the custom to apportion the power of a new vessel according to the formula

$$C = \frac{V^3 \times D^{\frac{2}{3}}}{I.\,H.\,P.},$$

the assumption being that the resistance varies as the square of the speed (V) and the power required to overcome it as the cube of the speed. Sometimes the area of midship section was introduced instead of the displacement (D), the resistance of ships of similar form but of different dimensions, being supposed to vary as the two-thirds power of those elements at any given speed, and the effective power directly as the indicated power, I. H. P.

Recent years have shown the fallacy of such assumptions, unless within narrow limits, and even then the forecast is not always to be trusted. We know from actual trial in America that it is possible to drive a larger displacement and midship section upon a decrease in beam with no greater expenditure in power. This has been sufficiently confirmed in the comparative sailing between "cutters" and "sloops" differing from one another in the respects noted, and is to be traced to the lesser wave-making of the narrower form. Trials of steam yachts in England have furthermore demonstrated that resistance does not vary as the square of the speed, but in a constantly growing ratio until high speed of 18 knots has been attained, when the ratio will slightly decline. It also varies with the form and dimensions. Neither does the I. H. P. vary as the cube of the speed. In the British yacht Mazeppa, for example, the I. H. P. from 6 to 8-knot speed increased as the square of the speed. From 8 to 10 knots as the cube, and from 10 to 11½ knots as the 4.4 power. The resistance of the same vessel, as measured by the dynamometer, grew directly as the speed up to 8 knots, from 8 to 10 knots as the square, and from 10 to 11½ nearly as the cube of the speed.

The readiest and most reliable method of apportioning power is, as with the screw to fall back upon experience and tentative progression.

From the foregoing it will be seen that every knot of increased speed is purchased at greater cost in power the higher the speed of the vessel, and that driving a hull beyond the speed to which it is economically adapted is a wasteful proceeding.

It is of course most economical to steam at slow speeds, as the distance covered per unit of fuel will be greater than at higher speeds with the power required growing in a faster ratio than the speed.

Thus a steam yacht of 500 tons displacement, carrying 100 tons of coal, upon trial was found to burn 2¾ tons per 24 hours at 8 knots, 5¼ tons at 10 knots, and 12¾ tons at 13 knots. The coal at these speeds would last nearly 36, 17 and 8 days, and the distance covered would be nearly 7,000, 4,100 and 2,400 miles. The economy per mile is only in the fuel. The longer time consumed in making a passage at low speed would involve other expenses as to subsistence, wages, etc., which might overbalance the saving in coal.

More specific information in further explanation of the foregoing general review will be found in the following chapters detailing actual practice.

VI.

U. S. LAWS APPLICABLE TO STEAM YACHTS.

THE exact status of a steam yacht in the eyes of the law has long been more or less indeterminate, depending upon such interpretations as were put upon the U. S. Revised Statutes by successive Secretaries of the Treasury and subordinate officials. Although legislation specifically exempts steam as well as sailing yachts from the burden of Custom House entry and clearance and from sundry taxes and dues, it has been generally held that the hull, equipment and machinery of a steam yacht are subject to the supervision of the Board of Supervising Inspectors of Steam Vessels and their Rules and Regulations established in accordance with the U. S. Statutes.

It is proper enough that the machinery of a steam yacht should by law be subjected to rigid inspection by disinterested government officers, both as a guarantee to the owner as well to the crew shipped at his instigation. Were such supervision removed, all manner of rash experiments would be indulged in with a train of serious consequences, and unscrupulous persons or over-zealous inventors would palm off upon the purchaser much worthless and dangerous material. There were hitherto some rules, which, though necessary and beneficial to public conveyances, were vexatious and not applicable to the proper and safe service of a small steam yacht. Among these were the licensing of master and pilot and the necessity of having such officers, the numerous articles of outfit for fire and life saving and the cost of inspection, which in proportion to the value of a small steam yacht was excessive.

Fortunately, the Treasury Department and the Board of Supervising Inspectors understand the situation well and have always been liberal in their interpretation of the Statutes. Amendments have been introduced and approved in many of the Rules, extending special privileges to yachts and vessels under one hundred tons. These have in a great measure relieved steam yachting from legal annoyances.

By Act of Congress, approved June 26, 1886, *all fees* for inspection of steamers and licensing of officers have been abolished. As the laws stand to-day, the machinery and hull of a yacht must be passed upon annually by the Local Inspectors for the District. Engineer, pilot or master's "special license" will be granted to any competent person upon application without payment of fee, but no unlicensed person can undertake their duties.

In small yachts and launches, under one hundred tons, as near as the law is specific on this point, no licensed "master" is required, but engineer and pilot of the smallest launch must have at least a "special license." It is well enough to insist upon a license for the engineer, but the pilot is a superfluous officer on board of any steam yacht. Where a licensed master is not required by law, the owner is sufficiently capable to act as his own "pilot," and in yachts large enough to require a "licensed master," the useless pilot can be dispensed with, for the master can in all cases also take out a license as pilot, if competent, thus combining the two officials in one person.

In 1878, Supervising Inspector-General Jas. A. Dumont caused the following circular from the Treasury Department to be issued. It covers the case of steam launches under the laws as they stand this day, with exception of the fee mentioned for "special engineer" license, which is no longer imposed.

U. S. Laws Applicable to Steam Yachts.

No. 2326, Synopsis of Decisions, 1875.

TREASURY DEPARTMENT, July 3, 1875.

[EXTRACT.]

* * * * * * * * *

Under Section 4426, Revised Statutes, the hull and boiler of every yacht, or other small craft of like character propelled by steam, must be inspected—the boiler being subjected to the hydrostatic test required by law. The pilot and engineer must also be licensed; and such other provisions of the law complied with as may be applicable to the particular vessel under examination.

Sections 4428 and 4431 require that the iron or steel plates of which the boiler is constructed must be stamped with the name of the manufacturer, the place where manufactured, and the number of pounds tensile strain it will bear to the sectional square inch.

The boiler must be provided with such appurtenances as are necessary to its safe management, viz: Feed pump and check valve, steam pressure gauge, safety valve, gauge cocks, a water gauge, (showing the height of the water in the boiler) and blow-off valve; and, if it is found applicable to the kind of boiler employed, a tin plug, so inserted that it will fuse by the heat of the fire when the water in the boiler falls below the prescribed limit.

There must be on board the means of applying the required hydrostatic test.

For so small a vessel as you describe (26 ft. long) four buckets kept on board will be sufficient means for the extinguishment of fire.

There must be provided for each person on board a life preserver containing at least six pounds of good block cork, adjustable to the body in the manner of a belt or jacket, with shoulder straps.

The fee for license as "special engineer" for this yacht, which will be granted to any person of good character, who has sufficient experience to manage the boiler and machinery safely, is five dollars. A similar "special license" as pilot for this vessel will be granted to any person of like good character who is familiar with the navigation in which she is to be employed, understands the Pilot Rules, and has had sufficient experience in handling this or other similar vessels.

U. S. Laws Applicable to Steam Yachts.

The master of a vessel of this class does not require license.

A steam whistle of suitable dimensions must be provided, with which the pilot will make the signals as required by the Pilot Rules above referred to.

When the equipment is completed and the vessel is ready for inspection, it is required that application shall be made in writing by the master or owner to the local inspectors within whose district the vessel is owned or employed.

*　　*　　*　　*　　*　　*　　*　　*　　*

Very respectfully,
[Signed]   CHAS. F. CONANT,
Acting Secretary.

The following extracts from the Rules and Regulations of the Board of Supervising Inspectors are instructive and have special interest to owners of steam yachts and direct bearing upon the lawful management of their vessels, after they are in possession. These "Revised Statutes" and "Rules" are in the amended shape up to January, 1886.

R. S. SEC. 4214. The Secretary of the Treasury may cause yachts used and employed exclusively as pleasure vessels, or designed as models of naval architecture, if built and owned in compliance with the provisions of sections forty-one hundred and thirty-three to forty-one hundred and thirty-five, to be licensed on terms which will authorize them to proceed from port to port of the United States, and by sea to foreign ports, without entering or clearing at the custom house; such license shall be in such form as the Secretary of the Treasury may prescribe. The owner of any such vessel, before taking out such license, shall give a bond in such form and for such amount as the Secretary of the Treasury shall prescribe, conditioned that the vessel shall not engage in any trade, nor in any way violate the revenue laws of the United States, and shall comply with the laws in all other respects. Such vessels, so enrolled and licensed, shall not be allowed to transport merchandise or carry passengers for pay. Such vessels shall have their name and port placed on some conspicuous portion of their hulls. Such vessels shall, in all respects, except as above, be subject to the laws of the United States, and shall be liable to seizure and forfeiture for any violation of the provisions of this title: *Provided*, That all charges for license and inspection fees for any pleasure vessel or yacht shall not exceed five dollars, and for admeasurement shall not exceed ten cents per ton. Fees abolished, Act of Congress, June 26, 1886.

SEC. 4426. The hull and boilers of every ferry boat, canal boat, yacht, or other small craft of like character, propelled by steam, shall be inspected under the provisions of this title. Such other provisions of law for the better security of life, as may be applicable to such vessels, shall by the regulations of the board of supervising inspectors, also be required to be complied with, before a certificate of inspection shall be granted; and no such vessel shall be navigated without a licensed engineer and a licensed pilot.

RULE I. Every iron or steel plate intended for the construction of boilers to be used on steam vessels shall be stamped by the manufacturer in the following manner: At the diagonal corners, at a distance of about four inches from the edges and at or near the center of the plate, with the name of the manufacturer, the place where manufactured, and the number of pounds tensile strain it will bear to the sectional square inch.

RULE I. 6. To ascertain the ductility and other lawful qualities, iron of 45,000 lbs. tensile strength, and under, shall show a contraction of area of 15 per cent., and each additional 100 lbs. tensile strength shall show one (1) per cent. additional contraction of area, up to and including 55,000 T. S. Iron of 55,000 T. S. and upward, showing twenty-five (25) per cent. reduction of area, shall be deemed to have the lawful ductility. All steel plate of one and one-half inch thickness and under shall show a contraction of area of not less than fifty (50) per cent. Steel plate over one-half inch in thickness shall show a reduction of not less than forty-five (45) per cent. *Provided, however*, That steel plate required for repairs to boilers built previous to April 1, 1886, may be used for such repairs when showing a contraction of area of not less than forty (40) per cent.

RULE II. 3. The pressure for any dimension of boilers must be ascertained by the following rule: Multiply one-sixth of the lowest tensile strength found stamped on any plate in the cylindrical shell by the thickness, expressed in inches or parts of an inch, of the thinnest plate in the same cylindrical shell, and divide by the radius or half diameter, also expressed in inches, and the sum will be the pressure allowable per square inch of surface for single-riveting, to which add twenty per centum for double-riveting.

RULE II. 4. The hydrostatic pressure applied must be in proportion of one hundred and fifty pounds to the square inch to one hundred pounds to the square inch of the steam pressure allowed

RULE II. 9. Lapwelded tubes shall have the following thickness in inches:

Diameter.	Thickness.	Diameter.	Thickness.
3	0.109	1¾	0.095
2½	0.109	1½	0.083
2¼	0.095	1¼	0.072
2	0.095	1	0.072

RULE II. 10. The strength of corrugated flues, when used for furnaces or steam chimneys, corrugation not less than 1½ in. deep, and provided that the plain parts at the ends do not exceed 6 in. in length, and the plates are not less than $\frac{7}{16}$ thick, when new corrugated and practically true circles, to be calculated from the following formula:

$$\frac{12,500}{D} \times T = \text{pressure,}$$

T being thickness in inches and D the mean diameter in inches.

RULE II. 11. The formulæ for cylindrical lapwelded and riveted flues in boilers to be used as furnaces, which shall be used in determining the pressure to be allowed shall be as follows:

Let D = diameter of flue in inches; A = 89,600, a "constant"; T = thickness of flue in decimals of an inch; L = length of flue in feet, not to exceed 8 ft.; P = pressure of steam allowable, in pounds.

$$P = \frac{89,600 \times T^2}{L \times D}$$

RULE II. 12. The feed water shall not be admitted into any boiler at a temperature less than one hundred degrees Fahrenheit for low-pressure boilers, and one hundred and eighty for high-pressure boilers.

13. Whenever steamers use a pressure upon their boilers exceeding sixty pounds to the square inch, they shall be inspected as high-pressure steamers and designated as such.

14. Vertical tubular boilers shall not be used on steamers navigating the Red River of the North and rivers whose waters flow into the Gulf of Mexico, unless the water-line is 2 in. above the upper end of the tubes and fire-line.

16. All steamers navigating the ocean, sounds, lakes, bays, and rivers, the boilers of which shall be internally heated, shall have a clear space of at least four inches on either side, and at the top not less than two inches clear space above the covering of the boilers.

17. All boilers hereafter placed in steamers shall have a clear space of at least 8 in. between the under side of the cylindrical shell and the floor or keelson.

All man-holes for the shell of boilers shall have an opening not less in diameter than 11 × 15 in. in the clear, except that boilers less than 34 in. diameter of shell have an opening in the clear, in man-holes of not less than 9 × 14⅞ in.; all boiler shells between 34 and 38 in. diameter, an opening not less than of 9 × 16 in., and all boiler shells between 38 and 48 in. in diameter, an opening not less than 11 × 15⅞ in.

18. All wood-work or other ignitible substance, approaching within 2 in. of the boiler, shall be suitably sheathed with metal, so adjusted as to permit a free circulation of air between the sheathing and the ignitible surface.

19. All boilers shall have a clear space at the back and ends thereof of 2 ft. opposite the back connection door. Slip joints in steam-pipes shall, in their working parts, when the steamer is to

be employed in navigating salt water, be made of copper or composition.

20. There shall be fastened to each boiler a plate containing the name of the manfacturer of the material, the place where manufactured, the tensile strength, the name of the builder of the boiler, when and where built.

21. Every sea-going steamer carrying passengers shall be supplied with an auxiliary or donkey boiler of sufficient capacity to work the fire-pumps.

22. All steamers shall have inserted in their boilers plugs of Banca tin, at least one-half inch in diameter at the smallest end of the internal opening, in the following manner, to wit: Cylinder boilers with flues shall have one plug inserted in one flue of each boiler; and also one plug inserted in the shell of each boiler from the inside, immediately before the fire line, and not less than four feet from the forward end of the boilers. All fire-box boilers shall have one plug inserted in the crown of the back connection, or in the highest fire service of the boiler. All upright tubular boilers used for marine purposes shall have a fusible plug inserted in one of the tubes at a point at least two inches below the lower gauge cock, and said plug may be placed in the upper head sheet when deemed advisable by the local inspectors. All fusible plugs, unless otherwise provided, shall have an external diameter not less than that of a one-inch gas or steam pipe screw-tap, except when such plugs shall be used in the tubes of upright boilers, plugs may be used with an external diameter of not less than that of a three-eighths of an inch gas or steam pipe screw-tap, said plugs to conform in construction with plugs now authorized to be used by this board; and it shall be the duty of the inspectors to see that these plugs are filled with Banca tin at each annual inspection.

23. All steamers having one or two boilers shall have three suitable gauge cocks in each boiler.

24. Lever safety valves to be attached to marine boilers shall have an area of not less than one square inch to two square feet of the grate surface in the boiler, and the seats of all such safety valves

U. S. Laws Applicable to Steam Yachts. 115

shall have an angle of inclination of 45 degrees to the center line of their axis.

Any spring-loaded safety valves constructed so as to give an increased lift by the operation of steam, after being raised from their seats, or any spring-loaded safety valve constructed in any other mannner or so as to give an effective area equal to that of the aforementioned spring-loaded safety valve, may be used in lieu of the common lever-weighted valve on all boilers on steam vessels, and all such spring-loaded safety valves shall be required to have an area of not less than one square inch to three square feet of grate surface of the boiler, and each spring-loaded valve shall be supplied with a lever that will raise the valve from its seat a distance of not less than that equal to one-eighth the diameter of the valve opening, and the seats of all such safety valves shall have an angle of inclination to the center line of their axis of 45 degrees. But in no case shall any spring-loaded safety valve be used in lieu of the lever-weighted safety valve without having been approved by the Board of Supervising Inspectors.

Donkey boilers used on all steam vessels for driving pumps, hoisting engines, electric lights or other purposes must be inspected the same as the main steam boilers.

The area of all openings in boilers and connections leading from boilers to safety valves shall not be less than the area of the valve used.

26. All boilers or sets of boilers shall have attached to them at least one gauge that will correctly indicate a pressure of steam equal to eighty per cent. of the hydrostatic pressure applied by the inspectors.

28. There must be means provided in all boilers using the low water gauges which are operated by means of a float inside the same, to prevent the float from getting into the steam pipe.

34. All holes cut through the bottom or bilge of a vessel that are covered by a sea valve or cock and secured to the skin of the vessel by bolts and connected to the engine and boilers by pipes, shall be arranged so as to be accessible at all times, so that if a leak or defect occurs it can be reached. Valves, seats, stems and bolts shall be of brass when used in salt water.

A stop cock or valve shall be placed between check valve and boiler on all feed pipes in order to facilitate access to connection.

RULE III. 2. The carrying capacity of all lifeboats shall be determined by the following rule: Multiply length, breadth and depth together and divide their product by ten; the quotient will be the number of persons such a boat is allowed to carry.

4. All lifeboats must have life-lines securely fastened to their gunwales, and a good rope painter of suitable size and length properly attached, and every lifeboat must be supplied with not less than four oars, and kept in good condition for immediate use.

9. All metallic lifeboats shall be furnished with an automatic plug.

10. One boat for steamers under 100 tons, two boats from 100 to 200 tons, three boats from 300 to 400 tons, five boats up to 500 tons, and six up to 1000 tons.

13. A portion of lifeboats may be replaced by approved life rafts.

14. All steamers navigating oceans, northwestern lakes and sounds, shall be equipped with life rafts in proportion of one to every two lifeboats required.

15. Rubber and canvas rafts to be kept inflated at all times. Granulated cork life rafts excluded from steamers.

17. Drags or floating anchors shall be constructed so as to be capable of being compactly stowed near the head of the ship. For ships of 400 tons or under, drag to have not less than 25 superficial feet. Steamers whose routes do not extend off anchorage are not required to have drags or floating anchors on board.

21. Every sea-going steamer and every steamer navigating the great northern and northwestern lakes carrying passengers shall not have less than three watertight cross bulkheads. Such bulkheads shall reach to the main deck in single decked vessels, otherwise to the deck next below the main deck. For wooden hulls they shall be fastened to suitable framework, which framework must be securely attached to the hull and calked. For iron hulls they shall be well secured to the framework of the hulls, and strengthened by stanchions of angle iron placed not more than two feet from center to center. One of the bulkheads must be placed forward and one abaft of the

engines and boilers. The third or collision bulkhead must be placed not nearer than five feet from the stem of the vessel. Iron bulkheads must be made not less than one-quarter of an inch in thickness, and wooden bulkheads must be of equal strength, and covered with iron plates not less than one-sixteenth of an inch in thickness.

23. All open steam launches or other steam vessels of five tons burden or less, carrying passengers, may dispense with the lifeboat when such vessels are provided with metallic air chambers placed under the seats and in the ends of said vessels of sufficient capacity to float the inert weight of said vessel, including her boilers and machinery, and such vessels shall also be provided with one life preserver for every person which the inspection certificate shall allow them to carry, including the officers and crew, and every such steam vessel, carrying fifteen passengers or less, shall carry at least two fire buckets and one axe.

RULE IV. 1. All passenger steamers are required to be provided with fire buckets, barrels and axes as follows: Steamers not over 50 tons, 5 buckets and 1 axe. Steamers not over 100 tons, 8 buckets and 2 axes. Steamers not over 200 tons, 1 barrel, 12 buckets and 2 axes. Steamers not over 500 tons, 2 barrels, 15 buckets and 3 axes. Buckets may be substituted for barrels. They should be filled with water.

8. All feed and steam pipes shall be attached at their terminal joints, with good and substantial flanges. Double-acting steam fire pump for steamer under 200 tons to have 4 in. stroke and 2 in. diameter of plunger. For steamer under 500 tons, 7 in. stroke and 4 in. diameter of plunger.

17. Steam siphon pumps which have been approved by the Board of Supervising Inspectors may be allowed in lieu of double-acting steam fire pumps on steamers under 100 tons.

RULE V. 1. Before an "original license" is issued to any person to act as a master, mate, pilot or engineer, he must personally appear before some local board or a supervising inspector for examination; but upon the renewal of such license, when the distance from any local board or supervising inspector is such as to put the person holding the same to great inconvenience and expense to appear in person, he may, upon taking the oath of office before any person authorized to adminster oaths, and forwarding the same, together with the license to be renewed, to the local board or the supervising inspector of the district in which he resides or is employed, have the same renewed by the said inspectors, if no valid reason to the contrary be known to them; and they shall attach such oath to the stub end of the license, which is to be retained on file in their office.

4. It shall be the duty of an engineer, when he assumes charge of the boilers and machinery of a steamer, to forthwith thoroughly examine the same, and if he finds any part thereof in bad condition, caused by neglect or inattention on the part of his predecessor, he shall immediately report the facts to the local inspectors of the district, who shall thereupon investigate the matter, and if the former engineer has been culpably derelict of duty, they shall suspend or revoke his license.

5. No "original license" shall be issued to any person to act as engineer, except for "special license" on small pleasure steamers, who cannot read or write, or who has not served at least three years in the engineer's department of a steam vessel, or as a regular machinist in a machine works, provided that any person who has served for a period of three years as a locomotive or a stationary engineer may be licensed to act as engineer on steam vessels after having not less than one year's experience in the engineer's department of a steam vessel.

6. The navigation of every steamer above 100 tons burden shall be under the control of a first-class pilot, and every such pilot shall be limited in his license to the particular service for which he is adapted. "Special pilots" may also be licensed for small steamers of all kinds, locally employed.

12. No original license for any route shall be issued to any person except for special license on small pleasure steamers and ferry boats navigating outside of ports of entry and delivery, who has not been employed in the deck department of a steamer or sail vessel for the term of at least three years preceding the application for license.

U. S. Laws Applicable to Steam Yachts. 119

14. Whenever a master desires to act as pilot, and furnishes the necessary evidence of his ability, the local inspectors for the district where the license is issued shall indorse the route on the master's certificate; and, in like manner, when a first-class pilot desires to act as mate, if the inspectors are satisfied of his ability, they shall indorse the fact on the pilot's certificate; *but a mate's license cannot be indorsed as first pilot.*

24. Starting, stopping and backing signals for steam vessels navigating the waters of the eighth and ninth supervising inspection districts:

Eighth district embraces all the waters of the lakes north and west of Lake Erie, with their tributaries, and the upper portion of the Illinois River down to and including Peoria, Ill.

Ninth district embraces all the waters of Lakes Erie, Ontario, Champlain, Memphremagog, and the river St. Lawrence, and their tributaries.

There shall be used between the master or pilot and engineer the following code of signals, to be made by bell or whistle, namely:

1 whistle or bell	. Go ahead.
1 whistle or bell	Stop.
2 whistles or bells	Back.
3 whistles or bells	Check.
1 long whistle or 4 bells	. Strong.
1 long whistle or 4 bells	All right.

Two whistles or 2 bells, when the engine is working ahead, will always be a signal to *stop and back strong.*

Masters and pilots of steamers on lakes and seaboard are required to have their wheel chains rove so that the wheel and helm shall move in the same direction, so that when the wheel is put to starboard the vessel's head shall go to port, and when the wheel is put to port the vessel's head shall go to the starboard.

RULE VI. 1. The annual inspection of a steamer must be made only on written application presented to the United States inspectors

U. S. Laws Applicable to Steam Yachts.

by the owner, master or authorized agent of the steamer to be inspected.

4. In the inspection of the hull of steamers, if the inspector shall not have satisfactory evidence otherwise of the soundness of the timber, he shall not give a certificate until the hull of the vessel shall be bored to his satisfaction.

RULE IX. 14. When it is known, or comes to the knowledge of the local inspectors, that any steam vessel is or has been carrying an excess of steam beyond that which is allowed by her certificate of inspection, it is recommended that the local inspectors in whose district said steamer is being navigated, in addition to reporting the fact to the United States District Attorney for prosecution under Section 4437, Revised Statutes, shall require the owner or owners of said steamer to place on the boiler of said steamer a lock-up safety valve that will prevent the carrying of an excess of steam, and shall be under the control of said local inspectors.

On the placing of a lock-up safety valve upon any boiler, it shall be the duty of the engineer in charge of same to blow, or cause the said valve to blow off steam at least once in each watch of six hours or less, to determine whether the valve is in working order, and it shall be his duty to report to the local inspectors any failure of such valve to operate.

In case no such report is made, and a safety valve is found that has been tampered with, or out of order, the license of the engineer having such boiler in charge shall be revoked.

RULE X. 2. All the equipments of a steamer, such as buckets, axes, boats, oars, rafts, shall be painted or branded with the name of the steamer to which they belong.

3. No oil that will stand a fire-test of less than 300 deg. Fahrenheit shall be used as stores on any steamer carrying passengers.

6. All steamers navigating rivers, lakes, bays and sounds in the night time, shall have one watchman at the bow.

PILOT RULES FOR LAKE AND SEABOARD.

RULE I. When steamers are approaching each other "head and head," or nearly so, it shall be the duty of each steamer to pass to the right, or port side of the other ; and the pilot of either steamer may be first in determining to pursue this course, and thereupon shall give, as a signal of his intention, one short and distinct blast of his steam-whistle, which the pilot of the other steamer shall answer promptly by a similar blast of his steam-whistle, thereupon such steamers shall pass to the right, or port side of each other. But if the course of such steamers is so far on the starboard of each other as not to be considered by pilots as meeting "head and head," or nearly so, the pilot so first deciding shall immediately give two short and distinct blasts of his steam-whistle, which the pilot of the other steamer shall answer promptly by two similar blasts of his steam-whistle, and they shall pass to the left, or on the starboard side, of each other.

NOTE.—In the night, steamers will be considered as meeting "head and head" so long as both the colored lights of each are in view of the other.

RULE II. When steamers are approaching each other in an oblique direction they shall pass to the right of each other, as if meeting "head and head," or nearly so, and the signals by whistle shall be given and answered promptly as in that case specified.

RULE III. If, when steamers are approaching each other, the pilot of either vessel fails to understand the course or intention of the other, whether from signals being given or answered erroneously, or from other causes, the pilot so in doubt shall immediately signify the same by giving several short and rapid blasts of the steam-whistle ; and if the vessels shall have approached within half a mile of each other, both shall be immediately slowed to a speed barely sufficient for steerage-way until the proper signals are given, answered, and understood, or until the vessels shall have passed each other.

RULE IV. When steamers are running in a fog or thick weather, it shall be the duty of the pilot to cause a long blast of the steam-whistle to be sounded at intervals not exceeding one minute.

Steamers, when DRIFTING or at ANCHOR, in the fair-way of other vessels in a fog or thick weather, shall *ring their bells* at intervals of not more than two minutes.

RULE V. Whenever a steamer is nearing a short bend or curve in the channel, where, from the height of the banks or other cause, a steamer approaching from the opposite direction cannot be seen for a distance of half a mile, the pilot of such steamer, when he shall have arrived within half a mile of such curve or bend, shall give a signal by one long blast of the steam-whistle, which signal shall be answered by a similar blast, given by the pilot of any approaching steamer that may be within hearing. Should such signal be so answered by a steamer upon the further side of such bend, then the usual signals for meeting and passing shall immediately be given and answered ; but if the first alarm signal of such pilot be not answered, he is to consider the channel clear and govern himself accordingly.

RULE VI. The signals, by the blowing of the steam-whistle, shall be given and answered by pilots, in compliance with these rules, not only when meeting "head and head," or nearly so, but at all times when passing or meeting at a distance within half a mile of each other, and whether passing to the starboard or port.

RULE VII. When two steamers are approaching the narrows known as "Hell Gate," on the East River, at New York, side by side, or nearly so, running in the same direction, the steamer on the right or starboard hand of the other (when approaching from the west), when they shall have arrived abreast of the north end of Blackwell's Island, shall have the right of way, and the steamer on the left or port side shall check her way and drop astern. In like case when two steamers are approaching from the east, and are abreast at Negro Point, the steamer on the right or starboard hand of the other shall have the right of way, and shall proceed on her

U. S. Laws Applicable to Steam Yachts. 123

course without interference, and the steamer on the port side of the other shall keep at a safe distance astern (not less than three lengths) until both steamers have passed through the difficult channel.

RULE VIII. When steamers are running in the same direction, and the pilot of the steamer which is astern shall desire to pass on the right or starboard hand of the steamer ahead, he shall give one short blast of the steam-whistle as a signal of such desire and intention, and shall put his helm to port; and the pilot of the steamer ahead shall answer by the same signal, or, if he prefer to keep on his course, he shall give two short and distinct blasts of the steam-whistle, and the boat wishing to pass must govern herself accordingly, but the boat ahead shall .in no case attempt to cross her bow or crowd upon her course.

N. B.—The foregoing rules are to be complied with in all cases except when steamers are navigating in a crowded channel, or in the vicinity of wharves; under such circumstances steamers must be run and managed with great caution, sounding the whistle, as may be necessary to guard against collision or other accidents.

SEC. 4233, REVISED STATUTES.—RULE XXIV. In construing and obeying these rules, due regard must be had to all dangers of navigation, and to any special circumstances which may exist in any particular case rendering a departure from them necessary in order to avoid immediate danger.

The line dividing jurisdiction between Pilot Rules on Western Rivers and Lakes and Seaboard at New Orleans shall be the lower limits of the city.

PILOT RULES FOR WESTERN RIVERS.

RULE I. When steamers are approaching each other from opposite directions, the signals for passing shall be one blast of the steam-whistle to pass to the right, and two blasts of the steam-whistle to pass to the left. The pilot on the ascending steamer shall be the first to indicate the side on which he desires to pass; but if the pilot of the descending steamer shall deem it dangerous to take the side indicated by the pilot of the ascending steamer, he shall at once indicate with his steam-whistle the side on which he desires to pass, and the pilot on the ascending steamer shall govern himself accordingly, the descending steamer being deemed to have the right of way. But in no case shall pilots on steamers attempt to pass each other until there has been a thorough understanding as to the side each steamer shall take. The signals for passing must be made, answered, and understood before the steamers have arrived at a distance of 800 yds. of each other.

RULE II. If from any cause the signals for passing are not made at the proper time, *as provided in Rule I.*, or should the signals be given and not properly understood, from any cause whatever, and either boat become imperiled thereby, the pilot on either steamer may be the first to sound the alarm or danger signal, which shall consist of *three or more* short blasts of the steam-whistle in quick succession. Whenever the danger signal is given the engines of *both steamers must be stopped* and backed until their headway has been fully checked, nor shall the engines of either steamer be again started ahead until the steamers can safely pass each other.

RULE III. When two boats are about to enter a narrow channel at the same time, the ascending boat shall be stopped below such channel until the descending boat shall have passed through it; but should two boats unavoidably meet in such channel, then it shall be the duty of the pilot of the ascending boat to make the proper signals, and when answered, the ascending boat shall lie as close as possible to the side of the channel the exchange of signals may have determined, as provided by Rule I., and either stop the engines or move them so as only to give the boat steerage-way, and the pilot of the descending boat shall cause his boat to be worked slowly until he has passed the ascending boat.

RULE IV. When a steamer is ascending and running close on a bar or shore, the pilot shall in no case attempt to cross the river when a descending boat shall be so near that it would be possible for a collision to ensue therefrom.

RULE V. When any steamer, whether ascending or descending, is nearing a short bend or point, where from any cause, a steamer approaching in an opposite direction cannot be seen at a distance of 600 yards, the pilot of such steamer, when he shall have arrived within 600 yards of that bend or point, shall give a signal of one long sound of his steam-whistle, as a notice to any steamer that may be approaching; and should there be any approaching steamer within hearing of such signal, it shall be the duty of the pilot thereof to answer such signal by one long sound of his steam-whistle, when both boats shall be navigated with the proper precautions, as required by preceding rules.

RULE VI. When a steamer is running in a fog or thick weather, it shall be the duty of the pilot to souud his steam-whistle at intervals not exceeding one minute.

RULE VII. When steamers are running in the same direction, and the pilot of the boat astern shall desire to pass either side of the boat ahead, he shall give the signal, as in Rule I., and the pilot of the boat ahead shall answer by the same signal, or if he prefer to keep on his course, he shall make the necessary signals, and the boat wishing to pass must govern herself accordingly; but the boat ahead shall in no case attempt to cross her bow or crowd upon her course.

RULE VIII. When boats are moving from their docks or berths, and other boats are liable to pass from any direction toward them, they shall give the same signal as in case of boats meeting at a bend; but immediately after clearing the berths so as to be fully in sight they shall be governed by Rule I.

RULE IX. All barges in tow of steamers between sunset and sunrise shall have their signal lights, as required by law, placed in a suitable manner on the starboard bow of the starboard barge, and on port bow of the port barge, which lights shall not be less than 10 ft. above the surface of the water.

LIGHTS FOR STEAM VESSELS.

STEAM AND SAIL VESSELS.

R. S. Sec. 4233.—Rule I. Every steam vessel which is under sail and not under steam shall be considered a sail vessel; and every steam vessel which is under steam, whether under sail or not, shall be considered a steam vessel.

LIGHTS.

Rule II. The lights mentioned in the following rules, and no others, shall be carried in all weather, between sunset and sunrise:

Rule III. All ocean-going steamers and steamers carrying sail, shall, when under way, carry:

(A) At the foremasthead, a bright white light, of such a character as to be visible on a dark night, with a clear atmosphere, at a distance of at least five miles, and so constructed as to show a uniform and unbroken light over an arc of the horizon of twenty points of the compass, and so fixed as to throw the light ten points on each side of the vessel, namely, from right ahead to two points abaft the beam on either side.

(B) On the starboard side, a green light, of such a character as to be visible on a dark night, with a clear atmosphere, at a distance of at least two miles, and so constructed as to show a uniform and unbroken light over an arc of the horizon of ten points of the compass, and so fixed as to throw the light from right ahead to two points abaft the beam on the starboard side.

(C) On the port side, a red light, of such a character as to be visible on a dark night, with a clear atmosphere, at a distance of at least two miles, and so constructed as to show a uniform and unbroken light over an arc of the horizon of ten points of the compass, and so fixed as to throw the light from right ahead to two points abaft the beam on the port side.

The green and red lights shall be fitted with inboard screens, projecting at least three feet forward from the lights, so as to prevent them from being seen across the bow.

U. S. Laws Applicable to Steam Yachts. 127

RULE IV. Steam vessels, when towing other vessels, shall carry two bright white masthead lights vertically, in addition to their side lights, so as to distinguish them from other steam vessels. Each of these masthead lights shall be of the same character and construction as the masthead lights prescribed by Rule III.

RULE V. All steam vessels, other than ocean-going steamers, and steamers carrying sail, shall, when under way, carry on the starboard and ports sides lights of the same character and construction and in the same position as are prescribed for side lights by Rule III., except in the case provided in Rule VI.

RULE VI. River steamers navigating waters flowing into the Gulf of Mexico and their tributaries shall carry the following lights, namely: One red light on the outboard side of the port smokepipe, and one green light on the outboard side of the starboard smokepipe. Such lights shall show both forward and abeam on their respective sides.

RULE VII. All coasting steam vessels, and steam vessels other than ferry boats and vessels otherwise expressly provided for, navigating the bays, lakes, rivers or other inland waters of the United States, except those mentioned in Rule VI., shall carry the red and green lights, as prescribed for ocean-going steamers; and, in addition thereto, a central range of two white lights; the after light being carried at an elevation of at least fifteen feet above the light at the head of the vessel. The head light shall be so constructed as to show a good light through twenty points of the compass, namely: From right ahead to two points abaft the beam on either side of the vessel; and the after light so as to show all around the horizon. The lights for ferry boats shall be regulated by such rules as the Board of Supervising Inspectors of Steam Vessels shall prescribe.

A bright white light, not exceeding twenty feet above the hull, shall be exhibited by all steamers when at anchor between sunset and sunrise, in a globular lantern of eight inches in diameter, so placed as to throw a good light all around the horizon.

Sailing vessels shall at all times, on the approach of any steamer during the night time, show a lighted torch upon that point or quarter to which such steamers shall be approaching. And upon any craft navigating rivers without being in tow of a steamer, such as rafts, flatboats, wood boats, and other like craft, they shall sound a fog horn at intervals of not more than two minutes.

It shall at all times be the duty of steamers to give to the sailing vessel, or other craft propelled by sails, every advantage, and keep out of her way.

In the case of the steam yacht Yosemite vs. the river steamboat Vanderbilt, the Court of Appeals decided in March, 1887, that steam yachts must show the light prescribed for coasting steamers under Rule VII.

VII.

EXTRACTS FROM LLOYD'S RULES.

CONCERNING STEAM YACHTS.

SEC. 6. Stern and propeller posts and after end of keel for single screw yachts must be double the sectional area prescribed for keels in the tables of the Yacht Register, and the keel tapered fair to it.

The portion of the forging of the stern frame, forming part of the keel, is to extend sufficiently forward for the after end of the scarph in paddle steamers to be at least once and a half the frame space before the sternpost; and in screw-propelled vessels at least twice and a half the frame space before the propeller post. The rudder braces to be forged on to the sternpost.

9. Steam vessels to have double reversed angle irons on floor under engines and boilers.

17. The garboard strakes of screw-propelled yachts, if seven-sixteenths of an inch, or more, in thickness, may be reduced one-sixteenth of an inch before the half-length only. Paddle steamers may also reduce the thickness abaft the half-length of the vessel. Boss plates covering the screw-shaft are to be of the same thickness as the strakes amidships of which they form part.

21. Steam vessels, where the number is 1500 and above, are to have iron watertight engine and boiler room bulkheads, and the space around the stern tube must be inclosed in an iron watertight compartment, and all vessels, when the number is 7000 and above, are to have an iron watertight bulkhead fitted forward. [These "numbers"

are obtained by adding half the breadth of the vessel amidships, the depth from top of keel to underside of upper deck beams and the girth of the half midship frame. The sum will give the "number" for regulating sizes of frames, reversed frames, floors and bulkheads. When multiplied by the length of the vessel, it will give the "number" for regulating sizes of keel, stem, post, keelson, stringer plates, side plating, deck tie plates and rudder.]

Watertight bulkheads must be fitted with sluice valves and cocks, to allow the bilge water to reach the pumps, and the valves must be controlled from above the waterline.

24. Skylights to engine rooms are in all cases to be substantially constructed; the coaming to which they are attached should be of iron, efficiently fastened to the beams. Skylights to be securely attached to the coamings and the glass in them should be protected, and in addition, deadlights must be fitted and arrangements made for their efficient security in bad weather.

26. Engine and boiler bearers must be properly constructed, and the engine seatings efficiently secured to them.

In vessels where the "number" is 3500 and above, double reversed angle irons must be fitted across the vessel to every floor under engines and boilers, and under the boilers the floor plates must be increased one-sixteenth of an inch in thickness.

27. Coal bunker openings are to be fitted with gratings and lids, and the lids must be secured with approved fastenings.

35. In cases where it is proposed to construct boilers of steel for classed vessels or vessels intended for classification, the material is required to fulfil the following conditions:

The material is to have an ultimate tensile strength of not less than 26 and not more than 30 tons per square inch of section, with an ultimate elongation of not less than 20 per cent. in a length of 8 in. It is to be capable of being bent to a curve of which the inner radius is not greater than one-and-a-half times the thickness of the plates or bars, after having been heated uniformly to a low cherry red and quenched in water of 82 deg. Fahrenheit.

Steel rivets are to be considered as part of the material, and in addition to being subjected to a shearing test, they must be capable

Extracts from Lloyd's Rules. 131

of withstanding the same tests as the plates are required to undergo.

All the holes in steel boilers should be drilled, but if they be punched the plates are to be afterward annealed.

All plates that are dished or flanged are to be annealed after the operations are completed.

No steel stays are to be welded.

Boilers to be tested to twice their working pressure.

Two safety valves to be fitted to each boiler, with combined areas at least half a square inch to each square foot of grate surface. Approved safety valve also to be fitted to the superheater.

Stop valves so that each boiler can be worked separately. Steam gauge to each boiler. Also blow-off cock independent of that on the vessel's outside plating.

Engines to have two feed pumps and two bilge pumps if over 70 H. P. Bilge injection must also be fitted to the circulating pump.

A donkey pump is to be provided for supplying boilers with water, taking it also from each compartment. If no hand pump is fitted, the donkey pump must be fitted to work by hand also.

Steam and feed pipes to be of copper.

Pipes through the bunkers to be protected.

Bilge suction to pump from each compartment.

Sea cocks to be above engine room platform. Gun metal rings around blow-offs.

The strength of circular shells to be calculated from the strength of the longitudinal joints by the following formula:

$$\frac{C \times T \times B}{D} = \text{working pressure.}$$

C = coefficient, as per table following. T = thickness of plates in inches. D = mean diameter of shell in inches. B = percentage of strength of joint found as follows, the least percentage to be taken:

$$\text{For plate at joint, } B = \frac{p-d}{p} \times 100.$$

$$\text{For rivets at joint, } B = \frac{n \times a}{p \times T} \times 100 \text{ with punched holes,}$$

$$\text{Or, } B = \frac{n \times a}{p \times T} \times 90 \text{ with drilled holes.}$$

Extracts from Lloyd's Rules.

In case of rivets being in double sheer, 1.75 a is to be used instead of a. In these formulæ, p = pitch of rivets; d = diameter of rivets; a = sectional area of rivets; n = number of rows of rivets.

TABLE OF COEFFICIENTS FOR IRON BOILERS.

	For ½ in. thickness and under.	For ½ to ¾ in.	Over ¾ in.
Lap joints, punched holes.......	155	165	170
Lap joints, drilled holes.........	170	180	190
Double butt strap, punched......	170	180	190
Double butt strap, drilled.......	180	190	200

TABLE OF COEFFICIENTS FOR STEEL BOILERS.

	¾ in. plates or less.	¾ to $\frac{9}{16}$ in.	Over $\frac{9}{16}$ in.
Lap joints....................	200	215	230
Double butt strap joints.........	215	230	250

For all shell plates of superheaters or steam chests, the coefficient should be two-thirds the above. All manholes to be stiffened with compensating rings. Shell plates under domes in boilers so fitted, to be stayed from top of dome or stiffened.

Stays supporting the flat surfaces are not to be subjected to a greater strain than 6,000 lbs. per sq. in. of section, or 8,000 lbs. if of steel, calculated from weakest part of the stay or fastening. No steel stays are to be welded.

The strength of flat plates supported by stays to be taken from the following formula:

$$\frac{C \times T^2}{P^2} = \text{working pressure in pounds per square inch.}$$

Where T = thickness of plate in sixteenths of an inch; P = greatest pitch in inches; C = 90 for plates $\frac{7}{16}$ thick and below, fitted with screw stays and riveted heads; C = 100 for plates above $\frac{7}{16}$; C = 110 for plates $\frac{7}{16}$ thick and under, fitted with screw stays and nuts; C = 120 for plates above $\frac{7}{16}$ thick; C = 140 for plates fitted with screw stays and double nuts; C = 160 for plates fitted with stays with double nuts and washers at least half thickness of plates and a diameter of $\frac{2}{3}$ of the pitch, riveted to the plates.

In case of front plate of boilers in the steam space, these numbers should be reduced 20 per cent., unless plates are guarded from direct influence of the heat.

The strength of girders supporting the tops of combustion chamber and other flat surfaces, to be taken from the following formula:

$$\frac{C \times d^2 \times T}{(L-P) \times D \times L} = \text{working pressure in pounds per square inch.}$$

Where L = length of the girder in inches; P = pitch of the stays; D = distance apart of the girders; d = depth of girder at center; T = thickness of girder at center; C = 6,000 if there is one stay to each girder, 9,000 if there are two or three stays, 10,200 if there are four stays.

The strength of furnaces to resist collapsing to be calculated from the following formula:

$$\frac{89,600 \times T^2}{L \times D} = \text{working pressure in pounds per square inch.}$$

Where 89,600 = constant; T = thickness of plate in inches; D = outside diameter of furnaces in inches; L = length of furnaces in feet.

If rings are fitted around the furnaces, the length to be taken between rings, and pressure not to exceed

$$\frac{8,000 \times T}{D}$$

The machinery and boilers of steam yachts are to be surveyed annually if practicable, and in addition to be submitted to a special survey every four years and the boilers to special survey when six years old and subsequently to annual survey.

If satisfactory, these surveys will be recorded in the Yacht Register thus:

"Lloyd's M. C. 5, 80" in red, or "B. & M. S. 5, 80" in red.

"Lloyd's M. C." (Lloyd's Machinery Certificate), denotes that the machinery and boilers are fitted in accordance with these Rules, and were found upon examination at the time to be in good condition.

"B. & M. S." (boilers and machinery surveyed) with a date, denotes that the boilers and machinery, though not fitted strictly in accordance with these Rules, were found upon inspection at that time to be in good condition.

"B. S." (boilers surveyed) with a date, denotes that the boilers were found upon inspection at that time to be in good condition.

SCANTLING FOR WOOD BUILT YACHTS.

Tonnage.	Spacing for Double Frames.	Floors Sided.	Frames sided and moulded at floors.	Frames moulded at deck.	Keel, stem post deadwood sided.	Outside planking.	Deck plank.	Sectional area of shelf.	Clamps, bilge streak.
	In.	In.	In.	In.	In.	In.	In.	In.	In.
20	18	4	3½	2¾	5½	1½	1¾	14	1½
50	20	5½	4½	3¼	7½	2	2¼	25	2
100	23	6½	5¾	3¾	9	2¼	2½	40	2¼
200	24	7¾	6¾	4¼	10	2½	2¾	55	2½
400	26	9½	8	5	11	3	3	75	3

Deck beams are proportioned according to the breadth of the vessel. For 8ft. beam, they should be 2½ in. square; for 12 ft., 4½ in.; for 16 ft., 6in.; for 20 ft., 7 in.; for 24 ft., 8 in., with reduction at ends. Spacing of beams to be 24, 30, 38, 42 and 46 in. respectively. A vessel of 20 tons should have 4 pairs of hanging knees to upper deck beams; one of 50 tons, 5 pair; one of 100 tons, 8 pair; one of 200 tons, 13 pair, and one of 400 tons, 18 pair.

Bolts through heel knee, deadwood, keelson, transoms, breasthooks and lower deck knees should be about $\frac{10}{16}$ in. for 20 tons; $\frac{12}{16}$ for 50 tons; $\frac{13}{16}$ for 100 tons; $\frac{14}{16}$ to 1 in. for 200 tons and $\frac{15}{16}$ to $1\frac{1}{16}$ for 400 tons.

Bolts through shelf, clamp and arms of knees should be about $\frac{8}{16}$ for 20 tons; $\frac{10}{16}$ for 50 tons; $\frac{11}{16}$ for 100 tons; $\frac{12}{16}$ for 200 tons and the same for 400 tons.

Extracts from Lloyd's Rules.

Bolts through heels of frames, upper deck shelf and clamp, and upper deck knees are of the same diameter.

Bilge and through butt bolts $1\frac{6}{8}$ for 20 tons; $1\frac{8}{8}$ for 100 tons; $1\frac{11}{8}$ for 200 tons and $1\frac{14}{8}$ for 400 tons.

Hardwood treenails should be 1 in. for 100 tons and $1\frac{1}{8}$ for 400 tons.

Steam yachts of 20 tons should have 2 anchors about 100 and 85 lbs. with at least $\frac{7}{16}$ studded chain or its equal in close-link. For 50 tons, there should be 3 anchors of 200, 150 and 84 lbs., with $1\frac{3}{8}$ chain. For 100 tons, 3 anchors, the largest about 350 lbs. with $1\frac{1}{2}$ chain; for 200 tons, 4 anchors, the largest about 525 lbs. with $1\frac{3}{8}$ chain; for 400 tons, 5 anchors, the largest 1,000 lbs. with $1\frac{1}{16}$ chain. Anchor stocks to weigh about one-quarter additional. Length of chain should run from 45 to 75 fathoms.

SCANTLING FOR IRON BUILT YACHTS.

No. for Frames, Reverse Frames and Bulkheads.	Angle Iron Frames.	Reverse Frames.	Floors.	Bulkheads.
20 to 22.5	$1\frac{1}{2} \times 1\frac{1}{2} \times \frac{4}{16}$	....	$7 \times \frac{4}{16}$	..
25.5 to 28	$2 \times 1\frac{3}{4} \times \frac{4}{16}$	$1\frac{1}{2} \times 1\frac{1}{2} \times \frac{4}{16}$	$9 \times \frac{4}{16}$	..
30 to 31.5	$2\frac{1}{4} \times 2\frac{1}{4} \times \frac{5}{16}$	$2 \times 1\frac{3}{4} \times \frac{4}{16}$	$11 \times \frac{4}{16}$	$\frac{5}{16}$
33.5 to 37	$2\frac{1}{2} \times 2\frac{1}{2} \times \frac{5}{16}$	$2\frac{1}{4} \times 2\frac{1}{4} \times \frac{5}{16}$	$12 \times \frac{5}{16}$	$\frac{5}{16}$
40.5 to 43.5	$3 \times 2\frac{1}{2} \times \frac{5}{16}$	$2\frac{1}{2} \times 2\frac{1}{4} \times \frac{5}{16}$	$14 \times \frac{5}{16}$	$\frac{5}{16}$
45.5 to 47.5	$3 \times 2\frac{3}{4} \times \frac{5}{16}$	$2\frac{1}{2} \times 2\frac{1}{2} \times \frac{5}{16}$	$15 \times \frac{5}{16}$	$\frac{5}{16}$
51 to 54	$3 \times 3 \times \frac{6}{16}$	$2\frac{1}{2} \times 2\frac{1}{2} \times \frac{5}{16}$	$16 \times \frac{6}{16}$	$\frac{6}{16}$

Numbers for keel, stem, post and plating.	Spacing of frames.	Keel, stem, sternpost.	Outside plating.	Keelson and Stringer angle irons.	Stringer plate or upper deck beams.	Thickness of wood-deck.	Diameter of rudder-head.
	In.		In.			In.	In.
900 to 1,200	18	$4\frac{1}{2} \times \frac{5}{8}$	$\frac{5}{16}$	$2 \times 2 \times \frac{4}{16}$	$8 \times \frac{5}{16}$	$1\frac{3}{4}$	$1\frac{1}{2}$
1,500 to 1,800	18	$4\frac{1}{2} \times \frac{7}{8}$	$\frac{6}{16}$	$2\frac{1}{2} \times 2\frac{1}{4} \times \frac{4}{16}$	$10 \times \frac{5}{16}$	$2\frac{1}{4}$	2
2,050 to 2,300	20	$5 \times \frac{7}{8}$	$\frac{6}{16}$	$2\frac{1}{2} \times 2\frac{1}{2} \times \frac{5}{16}$	$13 \times \frac{5}{16}$	$2\frac{1}{2}$	$2\frac{1}{2}$
2,600 to 3,100	20	$5\frac{1}{2} \times 1\frac{1}{8}$	$\frac{7}{16}$	$3 \times 2\frac{1}{2} \times \frac{5}{16}$	$15 \times \frac{5}{16}$	$2\frac{1}{2}$	$2\frac{3}{4}$
3,900 to 4,650	21	$6\frac{1}{4} \times 1\frac{1}{4}$	$\frac{7}{16}$	$3 \times 2\frac{1}{2} \times \frac{6}{16}$	$18 \times \frac{5}{16}$	$2\frac{1}{2}$	$3\frac{1}{4}$
5,350 to 6,000	22	$6\frac{3}{4} \times 1\frac{1}{4}$	$\frac{7}{16}$	$3\frac{1}{2} \times 2\frac{1}{2} \times \frac{6}{16}$	$22 \times \frac{5}{16}$	$2\frac{3}{4}$	$3\frac{3}{4}$
7,300 to 8,500	22	$7 \times 1\frac{1}{4}$	$\frac{8}{16}$	$3\frac{1}{2} \times 3 \times \frac{6}{16}$	$28 \times \frac{5}{16}$	3	$4\frac{1}{2}$

Middle line keelsons of plate iron from $\frac{5}{16}$ to $\frac{7}{16}$ thick are required in yachts whose number is above 3,100. Tie plates must also be worked over the beams. They are the same thickness as the stringer plates, but a little narrower. Garboard plates are $\frac{1}{32}$ thicker than side plating up to number 2,050 and $\frac{1}{16}$ thicker above that. Stem, post and keel to be double riveted when plating is over $\frac{1}{16}$ thick. Butts of plating and stringers double riveted when over $\frac{5}{16}$ thick. Upper edge of garboards and sheer strake double riveted over when $\frac{5}{16}$ thick.

Beams are proportioned to the breadth of the vessel. For 8 ft. breadth, the beams should be $2\frac{1}{2} \times 2 \times \frac{5}{16}$ angle iron; for 12 ft., they should be $3\frac{1}{2} \times 2\frac{1}{2} \times \frac{5}{16}$; for 16 ft., beams are $4\frac{1}{2} \times 3 \times \frac{5}{16}$; for 20 ft., they are of $5 \times \frac{7}{16}$ in. bulb iron and for 24 ft., of $6 \times \frac{7}{16}$ bulb iron.

Rivets must be their own diameter from end of plate at butts and the rows must be one and a half diameter apart in edge riveting. They should be spaced four and a half diameters apart from center to center in plating and from five to seven diameters apart in angle iron work. Rivets for $\frac{7}{16}$ iron are $\frac{1}{2}$ in. diameter; for $\frac{9}{16}$ iron, they are $\frac{5}{8}$ diameter and for $\frac{11}{16}$ they are $\frac{7}{8}$ in. No plates for the vessel's side to be less than five frame spaces in length. No butts of plating in adjoining strakes to be nearer than two spaces, and the butts of alternate strakes to shift not less than one frame space. All butts throughout the vessel should be given good shift, as a line of butts is a line of weakness.

VIII.

RACING STEAM YACHTS.

THE crudest method of racing steam yachts is to start all vessels at the same instant and note the times of arrival, making no allowances of any sort. The result determines which is absolutely the speediest yacht, without reference to the means by which the speed has been produced or the possibilities for speed possessed by vessels differing in hull and driving power. Such tests have interest for a time, but would not promote competition nor furnish any desirable information concerning the economics of form in hull or engine performance. It would soon become evident that the longest or largest yacht would have an undue advantage from the greater possibilities inherent in mere size, providing always that due advantage has been taken of the possibilities by the designing engineer. The entries would rapidly diminish, as none but the largest would have any chance in such competition. The races would leave us none the wiser in the end.

If, however, a shorter or smaller vessel should demonstrate herself absolutely faster than a longer competitor, she would receive credit therefor in her victory. But the amount by which she actually beat the larger boat would not truly represent her superiority, for the victory was gained upon a less length, and therefore upon smaller possibilities so far as *hull* is concerned. An allowance for this difference in length of the competitors would in justice have to be added, so that the absolute amount by which the smaller yacht distanced her larger rival, may be augmented in proportion to her lack of length.

It is upon this consideration that the "Emory Tables" have been compiled. Chief Engineer F. B. Isherwood, U. S. N., counselled an allowance based upon the assumption that speeds will or should vary as the cube root of the loadline lengths, and Mr. C. E. Emory found that by multiplying the cube root of the length by 2.7 it would very nearly express the actual speed attained by yachts of normal form and power, or $2.7 \sqrt[3]{L}$

TIME ALLOWANCE FOR LENGTH BY EMORY FORMULA.

Length of loadline.	Time required to run 80 knots.	Length of loadline.	Time required to run 80 knots.
Feet.	H. M. S.	Feet.	H. M. S.
50	8 02 33	180	5 17 51
55	7 47 28	185	5 14 53
60	7 33 42	190	5 12 00
65	7 21 32	195	5 09 14
70	7 10 57	200	5 06 35
75	7 01 38	205	5 03 59
80	6 51 49	210	5 01 32
85	6 44 49	215	4 59 05
90	6 36 42	220	4 56 44
95	6 29 53	225	4 54 29
100	6 23 01	230	4 52 19
105	6 16 50	235	4 50 10
110	6 11 02	240	4 48 06
115	6 05 33	245	4 46 05
120	6 00 25	250	4 44 07
125	5 55 40	255	4 42 12
130	5 50 55	260	4 38 33
135	5 46 34	265	.. .. ..
140	5 42 15	270	4 35 05
145	5 38 21	275	.. .. ..
150	5 34 36	280	4 31 47
155	5 30 58	285	.. .. ..
160	5 27 27	290	4 28 36
165	5 24 09	295	.. .. ..
170	5 20 53	300	4 25 35
175	5 17 51		

The foregoing allowance is really only one for the hull, and would be fair and all that is required upon the assumption that in every steam yacht built, every quality is sacrificed to the sole attainment of the highest speed. But this is manifestly not the case, and the assumption would in the long run be fraught with evil consequences to the sport. In the Emory Rule, it is assumed that upon any given length, the motive power will take up the same proportion of the displacement, and that an addition to displacement means additional opportunity for increasing weight of motive power. These are two suppositions which will not always hold.

In the first place, many steam yachts are intended to meet other requirements than the maximum speed only. They will differ in strength of construction, according to the service they are to undergo. It would be unwarranted and reckless to construct a seagoing steamer in the same light and trifling way that high speed launches and smooth-water craft are produced. More solid structure, more freeboard, stores, coal bunker capacity, rig, equipment and varying demands as to accommodations, necessarily destroy the correctness of the assumption in the Emory Rule, that motive power will bear the same proportion to displacement in all yachts of like length. Economy in space taken up by engines and boilers and small coal consumption, which are of great importance to cruising yachts, will frequently place another limit upon the motive power, as also considerations of first cost and subsequent running. Thus it will be seen that individual yachts will not necessarily meet the conditional characteristics upon which the equity of the Emory Rule rests.

The direct tendency of the Rule will be to banish from the start all but avowed "racing machines," and in these the weight of structure and equipment will be reduced to a minimum, rendering the vessels unreliable, if not actually unsafe, and certainly unfit to meet the needs of the seagoing cruiser. Yachts of normal proportions, normal power and high performance as to engine economy would not be countenanced under the Emory Rule. Nor would efforts in behalf of improvement in performance, which are at least equally as important as the maximum possible speed on the length, be given any consideration. So far from furthering advance in engineering questions, a "length rule" would rather promote the most costly and wasteful machinery, the "speed-at-any-price" style of yacht, poorly suited to the much wider field which legitimate steam yachting should really command.

In the Emory Rule, the competitive equity must be deemed as carried out only half way. If two yachts of like length meet and race over a course in equal time, they would pass off as equally efficient under the "length rule." Yet one of the yachts may have accomplished the distance upon less expenditure of power, showing that her model or machinery is in reality the superior of the two. In equity, and for an intelligent appreciation of the competition, this superiority should be made evident in a tangible way for true comparison of performance, otherwise the relative perfection of model and machinery between the two yachts would be lost sight of. Evidently an allowance for the smaller expenditure of power is a proper correction to apply, thereby ascribing to the boat expending least power that superiority in performance which is really hers.

The Emory rule is also lacking in other material respects. If two yachts of like length are driven by like power, it stands to reason that a fine form will have greater possibilities from the very start than a fuller form of hull, as the extra displacement of the latter does not necessarily go to greater driving power, and in this example may be supposed to be expended in a structure of sea-going scantling and in storage for distant cruising. Manifestly, such a vessel should in equity be entitled to an accounting for the greater bulk or volume she has driven through the water with like speed and like power. Her performance is certainly more creditable under these suppositions than that of her finer lined opponent. Yet competition under "length measurement" would rob the fuller boat of the credit she is justly entitled to, and in so doing would fail to disclose a true understanding of the respective merits of the two vessels.

If the racing of steam yachts is to be viewed simply as a rough lumping together of vessels of various sizes (lengths) with a view to distributing prizes to the first boat across the finish, regardless of the real merits of performance, the Emory Rule will be sufficient, though it will confine racing to a few "machines" when its working is fully comprehended.

But if racing steam yachts is to be placed upon a higher plane of justice and engineering utility, and permit comparative tests between all varieties without prejudice to some and a premium upon others, then it is manifest that "length measurement" must be replaced by a rule which shall take due consideration of displacement and power as well as length. The length must still be included, for it is well known that a given displacement in a long form has greater possibilities for speed than if placed in a shorter, chubbier form, at least within such proportions as have been practically exploited at this day.

If one vessel has a larger displacement than another of the same length, and that excess of displacement has been utilized in greater weight of engine and more driving power, the excess of power expended over the course will pay under the proposed rule its just share for the greater possibilities possessed in the excess of displacement as expressed in the increased power.

Mr. Chas. H. Haswell, Measurer of the American Yacht Club, has contributed a rule which comes near to satisfying the above train of reasoning and also seeks to cover the features of natural and forced draft. Mr. Haswell accepts the old formula of naval architecture that speed will be proportional to the cube root of the power divided by the area of the midship section, or else by the two-thirds power of the displacement. The length receives no consideration. Yet of two vessels having the same midship section, the longer has a fundamental advantage, and the same can be said of displacement. Hence, the introduction of length as a factor in a just formula is a necessity.

It will not do to say it is the business of the builder to distribute this displacement or cross-section upon the most advantageous length, for such proportions would lead to the construction of "machines" and drive out of existence the normal steam yacht upon the propagation of which the popularity and usefulness of steam yachting mainly depend.

The designer, who from other considerations than speed, is forced to choose a less favorable length to the displacement or cross-section, should not be "boycotted" from the line, for his production may in fact be more creditable from an engineer's or yachtsman's standpoint, though not as fast as the "machine" when compensation is given under a rule *prejudicial* to one and *favorable* to another *regardless of intrinsic merit of performance.*

Mr. Haswell explains his method of determining allowances as follows :

The end desired, that is, a comparison of volume, can be attained with all sufficient accuracy and with more facility by taking the gross Custom House tonnage, as computed either by United States or British laws, which is a fair exponent of the volume of the hull of a vessel.

To arrive at the other element, that of actual power, by the computation of it from an engine by its operation, involves time and much labor, in addition to the difficulty of deriving reliable assistance from the driver of the engine to develop the full power of the boiler, as shown by the operation of the engine, when the development of it is to operate to the disadvantage of his employer.

The determination of the capacity or power of a boiler would be best attained by ascertaining the exact volume or weight of water evaporated, or fuel consumed in an assigned period ; but inasmuch as to attain these results would involve expensive attachments to a boiler, labor and time, and be exposed to the objections already referred to in relation to the engine driver and his assistants in interest of their employer, both of these methods would seem to be impracticable of adoption.

In order, then, to exclude all elements of variation, all dependence upon the integrity of operatives, and to operate only with such elements as combine the greatest uniformity, facility and practicability of attainment, and assuming that the speed of similar vessels is as the cube root of their moments, I submit as follows :

The accepted formulas for the speed of steam vessels :

$$\sqrt[3]{\frac{P}{A}} \text{ and } \sqrt[3]{\frac{P}{D^{2/3}}} \text{ each } = V.$$

P representing the horse power; A area of immersed amidship section, in square feet; D displacement of hull in tons ; and V velocity of vessel in miles per hour; the former when the area of amidship section is taken, and the latter when the displacement of the hull is taken.

The boiler, that is the area of the grate surface, and the character of its construction and combustion, are the essential elements of the power of a steam vessel. The dimensions of the engine are arbitrary and secondary.

Thus, with like and equal boilers, the attached engines may be of different diameters and stroke, the propellers of different areas and pitch of blades, yet the power under like combustion is constant. The manner of utilizing it by high or low expansion, by high or low velocity of piston and of propeller, is arbitrary; and in the attainment of the best results with the least means, is the field of competition; as in the manner of two sailing vessels of like designs and similar capacities, one may be rigged taunt, with light canvas, the other square, with heavy canvas, the competition would be in arriving at that sparring, rig and fitting which is most effective.

If the manner of combustion in steam boilers was uniform, computation of their power as determined by the area of their grate would be very simple, but, inasmuch as there are four manners of operating them, it becomes necessary to assign a specific rate or factor for each; thus, there is combustion by natural draft, steam jet, blast and exhaust, whereby the effect or capacity of grate surface is materially altered, and if their relative effects can be arrived at with any reasonable accuracy, the problem of relative capacities of steam yachts is very satisfactorily attained.

From experiments lately conducted in England with boilers of two steamers, to determine the relative effects of different manners of combustion, the results were as follows:

Natural draft, 1; jet, 1.25; blast, 1.6. Adopting these figures as fair exponents of the case, the grate surface of a yacht then should be multiplied by the factor due to the manner of combustion to arrive at the power, and the volume of the vessel in tons, being taken to represent her mass, it remains then only to reduce these elements to a standard of comparison in order to assign a just allowance of time for the difference of the elements; and in order to arrive at this, and assuming the second formula above given, viz., that the speed of a steam vessel is as the cube root of the square of her displacement in tons, then for power of engine substitute area of grate, due to the method of combustion, and it remains but to decide upon the allowance of time for the greater capacity to give the less in each class, and which results may be satisfactorily arrived at by a summary of the elements and speed of a number of vessels.

In illustration of this method, I submit two yachts, having capacities and powers as follows:

No. 1. 80 tons, 40 sq. ft. of grate, with a blast draft, $= 40 \times 1.6 = 64$ sq. ft. Then

$$\sqrt[3]{\frac{64}{80^{2/3}}} = \sqrt[3]{\frac{64}{18.57}} = \sqrt[3]{3.447} = 1.511$$

No. 2. 70 tons, 50 sq. ft. of grate, natural draft, $= 50$. And

$$\sqrt[3]{\frac{50}{70^{2/3}}} = \sqrt[3]{\frac{50}{16.98}} = \sqrt[3]{2.94} = 1.433.$$

Which results would represent their competitive capacities. Assuming then the speed of No. 1 to be 14 knots per hour

$$= \frac{6,120 \times 14}{60} = 1,428 \text{ ft. per minute};$$

and that No. 2 will be as $1.511 : 1.433 :: 1,428 : 1,354$ ft.

Then, if 1,354 ft. require one minute, 74 (1,428 — 1,354) will require .05466 minute more time to attain a like distance, $= .05466 \times 60 = 3.2796$ minutes per hour, and

$$\frac{.05466}{14} = .003904 \text{ per minute.}$$

Hence, for course of 50 nautical miles, the allowance No. 1 would have to give No. 2, would be $.003904 \times 60$ (minutes per hour), and by 50 (miles) $= 11.712$ minutes.

Inversely: $6,120 \times 50 = 306,000$ ft., and

$$\frac{306,000}{1,354} = 214.3 \text{ minutes,}$$

the time of No. 1 performing the distance, and

$$\frac{306,000}{1,354} = 225.9 \text{ minutes,}$$

the time of No. 2 performing the distance. Hence, $225.9 - 214.3$

Racing Steam Yachts.

= 11.6 minutes, difference or allowance to be given for a distance of 50 miles.

In the application of the first formula to attain V or speed of vessel, it is usual to add a coefficient to the numerator, which represents the relative capacity of performance.

In the case under consideration it is not the velocity that is required, but the elements that produce it, and the coefficient is the representation of the competitive capacity of the yacht. Thus, the finer the model the more effective the instrument of propulsion, as a propeller or side-wheels, and the less the friction, both of the engine and the wet surface of the hull, the greater is the coefficient; or, in other words, the more effective the result attained, and as this coefficient is peculiar to each and every vessel, it is of value only when known, and consequently, except in a general application to like types and like proportions of vessels, it is useless.

IX.

MANAGEMENT AND CARE OF MACHINERY.

FILLING UP THE BOILER.—Cold water pumped into hot boilers is very injurious, causing severe contraction of the seams and stays, which very often leads to fracture of the stays or leakage in the seams and tubes. Many boilers, well constructed and of good material, have been ruined by being blown out under a high pressure of steam, and then immediately filled with cold water. The boiler should be allowed to cool down first. A mark on outside of boiler and on the water glass should indicate the height of the crownsheet, so that the fireman may be certain of carrying the water level high enough to avoid exposure. If no such mark has been made, fill up the boiler and with the manhole open make the necessary observation and transfer the measurement to outside of boiler and from that to the water glass.

LIGHTING FIRES.—Fires should not be started until the boiler has been pumped full. If too much water has been pumped in, blow off by the bottom valve, after the water is partly heated. This will withdraw cold water from the bottom and start the circulation. Start the fires in ample time and do not force them with cold water in the boiler. The grate should be kept well covered with a thin fire. Do not feed with large lumps or too much at a time, or keep the fire door open too long. Keep the grate free from clinker, so that the draft may not be impaired. The fires are started by splitting a quantity of wood and distributing it with shavings and oily cotton waste over the grate. When this has reached a blaze and the ashes glow, introduce a little fine coal without smothering the wood fire. Cotton waste should not be kept stored up for this purpose, however, as it is liable to spontaneous combustion.

A slow rate of cumbustion with moderate draft produces a better evaporative result than when the fires are urged. The test of a good fire is in the glow of the ashpit. When the ashes in the pit appear dull, the fire needs cleaning, but it should not be broken up. The bars should be evenly covered and no space left bare, as a cold current of air would draw up and sweep over the fire, cooling it down. In boilers with two furnaces attend one at a time. Bituminous coal is apt to form a crust on the surface, and before feeding should be broken. Avoid heaping the fuel at the front, and do not clean a fire down too low, as it will take some time to come up again.

SAFETY VALVE.—Raise the safety valve to permit hot air to escape. When a few pounds pressure are shown on the gauge, open stop and throttle valve and allow steam to pass through the engine to warm up its parts. Sudden admission of high steam to the cold engine would cause such expansion to packing of piston and other light parts that free working will be endangered. When steam is up, the fires should be so managed that the safety valve will not blow off, although the point of blow-off should first be compared with the steam gauge.

If the boiler steams too fast, close the damper and shut off draft, but do not throw open the furnace door if it can be avoided.

FOAMING OR PRIMING.—This is the violent ebullition of the water at the surface, and is caused by irregular feed, sudden withdrawal of steam, as upon opening the throttle wide at once, or from dirty water. Also from the presence of grease, especially in new boilers. Most frequently the trouble is with the manner of feeding. To avoid taking foam over into the steam pipe, the under side of the dome is protected by perforated plate, or the supply pipe is extended the length of the boiler inside and perforated, so as to take the steam from the whole length and not from one spot. Wash plates are also sometimes fitted in the boiler near water level, and a trap on the pipe between boiler and engine, to catch the watery particles before reaching the engine. To check priming, close throttle valve long enough to show true level of water. If sufficiently high, blow from the surface and turn on feed. In case of violent foaming caused by dirty water, or change from fresh to salt or the opposite, check draft in addition and cover the fires with fresh coal. In other words, cool down the fire so that the ebullition may cease. If grease is the cause, common washing soda pumped into the boiler will stop it. Feed water should not be taken from greasy tank or oil barrel. Carrying the water too high is another cause of priming, as it reduces the steam space. Boilers with insufficient steam space are liable to frequent priming.

THE FEED.—The working of the feed can be followed by the rise in the water glass. If there are doubts about it, feel the pipe near the check valve. If it warms up, the check is at fault, for back water is escaping from the boiler. By placing the ear close to the valve, the click of the valve can be detected. If it is working, look about for some source of escape in the boiler. Examine the blow-off cocks and other connections to ascertain if they are closed. Then turn to the piping and look for a leak in the pump delivery or suction, or the stopping up of the latter. A burst can be temporarily repaired by wrapping it with canvas coated with white lead and serving it over with marline or twine. For a long split lay a strip of wood or iron on before applying the canvas. The feed pump should not be allowed to heat up from too close proximity to the boiler, or from undue friction of working parts, or from drawing the suction too hot from the hot-well. A heated pump will generate vapors which prevent the valves from working free in the plunger cylinder.

LOW WATER.—Should the water meanwhile run down in the boiler, draw the fire and leave furnace door open, or bank the fires or cover with fresh coal and leave furnace door open. But if the crown sheets are supposed to be overheated, cover the fire with damp ashes, open door and close the damper. Under no circumstances turn on the feed full speed and force cold water over the hot sheets, should it suddenly be found to work. This is a frequent cause of explosions. A very slow feed may be kept up, if it has been going at the time, but if engine and feed have been stopped do not start them, as a sudden commotion in the boiler might cause disaster. If the engines are going, do not interfere, and do not suddenly lift the safety valve, as the boiler should be left at rest under the circumstances and not be submitted to any violent disturbance. When the boiler has cooled a little, the safety valve may be lifted gently and the engines started to run down the steam. The fire should not be drawn in the extreme case of hot crown sheet, because much heat would be liberated during the attempt. It is better to smother it and open the smoke-box door to draw cold air through the tubes. In hastily drawing fires, the hot coal is dumped in front of the boiler, and the fireman finds he cannot complete the work, owing to the heat rising from the pile. He is then in trouble and apt to lose his head and quit his post, demoralizing those around him.

After steam has run down, the boiler must be emptied and examined by a person capable of estimating the damage before risking a fresh start.

With proper gauge, cocks and low water alarm, frequently tested, only gross negligence or incompetency can bring about overheated crown sheets, and even then the fusible plug should avert danger if it has been properly attended to.

INSPIRATOR FAILS TO FEED.—If the steam pipe is full of hot water when the Hancock Inspirator is to be started, open the steam valve sufficiently to allow the water to pass off through the inspirator and out at the overflow. If the steam pipe connecting with boiler is long, it should be of larger size than the inspirator connections, and if the pipe is horizontal, pitch it so as to return condensation back to boiler.

If the suction is filled with hot water from the same cause, it may be remedied in two ways. Cool the inspirator and suction with cold water, or else pump the water out by letting the steam on and off suddenly at the starting valve of the instrument, until the hot water has been disposed of.

If the inspirator does not lift water well the difficulty will nearly always be found with the suction, which must be absolutely tight to secure good results. The lift should not be out of proportion to the steam pressure, the overflow should be wide open and not choked by being piped too small. The steam and air should be given free vent at the overflow to raise the water. Sometimes when the suction is very hot, the water becomes very much heated by the time it reaches the inspirator, and it will not condense the steam; the water will come up into the inspirator, but will not pass through the jet. The simplest way to overcome this trouble is to shut off the steam and let the water down. This will cool the inspirator and suction and you can then let on the steam again and get the water without difficulty. If, after the water is got, it will not penetrate to the boiler, the cause is often due to "starving" the inspirator, not giving it water enough. It is caused by having the suction too small, so that the "lifter" does not supply the "forcer" with water enough to condense the steam on the forcer side, hence the inspirator refuses to work. Sometimes, owing to a leaky steam valve, the first water that comes is very hot, and then the forcer cannot take care of it, as it will not condense the steam. When this is the trouble, let the water run out at the overflow until it becomes cooler, then the forcer will take it and send it to the boiler. See that the check valve in the delivery pipe to boiler is not stuck down, and that it has enough play not to choke the delivery. This pipe should be as large as the inspirator connections. A leaky suction will also prevent the water going to the boiler when the forcer valve is opened and the final overflow closed.

BLOWING OFF OR BLOWING DOWN.—Boilers should never be blown out while hot, as the plates and tubes retain sufficient heat to bake the deposits into a hard scale that becomes firmly attached to their surface. The boiler should always be allowed to cool down before the water is run out; the deposit will then be quite soft, and can be washed off with a hose. Many engineers suppose that blowing out under pressure tends to force out these deposits from the boiler, but experience has shown this to be a mistake.

BANKING FIRES.—During stoppages of some duration, the fires are "banked" to preserve them for ready use without consuming fuel.

They are simply smothered partially with wet slack and ashes, the damper is closed, furnace door opened and smoke-box door opened. A little attendance to the draft will keep the fires at a standstill. If steam begins to blow off, turn on cold feed.

TO REMOVE SEDIMENT AND DEPOSIT.—The handholes should be frequently removed and all incrustation and dirt cleared out. Plates exposed to the fire should be kept clean, also the tubes, flues and connections. If the tube plates and tubes are heavily coated with scale, the slice bar and hammer must be brought into use. The scale can be loosened by injecting 1½ oz. of muriate of ammonia twice a week, which will materially assist. On no account should the boiler be heated and then filled with cold water with the object of cracking the scale off through sudden contraction of the metal. Scale should not be allowed to accumulate to a greater thickness than that of a sheet of writing paper. The tubes are swept out by wire brushes, the soot being moistened as it comes out. Or they may be blown out by a jet of steam from a strong hose with a nozzle.

Internal or external corrosion should be prevented. If any pitting is observed inside, a coating of Portland cement mixed with litharge and linseed oil should be applied as a protection to the iron. External corrosion will most frequently start in places where the boiler is not readily accessible. Periodical inspection with painting is the only preventive.

SPLIT TUBES.—Should a tube start leaking when under steam, it must be plugged. Wood plugs can be driven in front and beyond the leak. They will swell tight and withstand considerable pressure, but may be blown out, when the fires will be quenched and the steam escape, which is often attended by serious results to those in the stoke hole. Regular tube stoppers should always be on hand. They consist of a rod as long as the tube, with washers and nuts at each end. The rod is pointed through the tube and the nuts set up This may involve drawing the fires and cooling down, according to the style of boiler. Canvas washers steeped in red lead are introduced under the iron washers to make a tight job.

BLISTERS.—When a blister appears, there must be no delay in having it carefully examined and patched.

GENERAL CARE.—In very shoal water slow down engine and feed pump so as not to suck grit from the bottom. A little of it under the valves of the pump will prevent its proper working. Try the pump occasionally by the water or pet cocks to see that it is acting. If it can be done, stop the feed until deeper water is reached. Keep an eye on water cocks and glass gauge and try them frequently to be certain of their free working. Keep the boiler dry when not in use by burning a few sticks of wood in the furnace, but not enough to heat up the shell. In open launches, a hood should be provided when the launch is not running.

WHEN LAID UP.—Dry the boiler, take off manhole covers, remove safety valve and cover the smokestack. Whitewash inside after cleaning. Uncouple valves and plug pipes with wood steeped in tallow and whitelead. Allow no water to remain in the elbows. Store away all attachments and take a look at the boiler occasionally, so that no water may collect and cause oxidation.

GETTING UNDERWAY.—See that all the manholes and handhole plates are fully screwed up over the packing grummets and open the blow-off cocks, allowing the water to run up as high as it will, unless it is proposed to fill up with fresh water from a hose on the dock. Meanwhile get the fires ready and start when the boiler has been pumped up by hand to the level required. Blow-offs are to be closed after running up the water, and the feed pump connected for working by steam after boiler is full, the suction opened and the check valve to suit as well as intermediate valves in the feed pipe. Injection and outboard delivery must be opened ready for the condenser to operate.

Open the stop valve slowly when steam has been raised. Then open the throttle valve partly and the blow-through valves and drain cocks on the engine. Allow the steam to drive the water and air from the steam chests, cylinders and passages, until steam alone issues from the drain cocks. Close the blow-through valves, and the steam having found its way into the condenser, will be precipitated by the cold injection water in the pipes and form a partial vacuum. Then throw the valve gear partly into action and start the engine slowly, being sure that there is nothing to foul the screw. The links may have to be thrown back full distance in some engines to give a good supply of steam to start the machinery from rest. But as soon as started, the supply should be checked by the throttle, which can then be gradually opened wide as the vessel moves away. All the machinery will then be in regular action. The circulating pump is forcing cold injection through the condenser pipes and the air pump is drawing off the condensed steam to the hot-well or tank from which the feed pump returns it to the boiler, if operated from the main engine, or else must be given steam independently to suit. The valves can be regulated by watching the glass water gauge.

In compound engines, a starting valve is placed on a connection between the steam chest of the high-pressure cylinder and the low-pressure cylinder, so that high steam may be admitted into the latter to assist in starting.

ATTENDANCE WHILE RUNNING.—See that the oil cups are full and wicks in order, and attend those of the cylinders, by closing the upper and opening the lower cock on the vacuum side of the piston. Watch the bearings that they may not heat. If this is found to be the case, slow down and turn on cold water to cool them. The stream should be played over the shaft and not over the brasses, as the latter are liable to crack or scale upon sudden cooling. A common cause of hot bearings is carelessness when polishing the engines with fine emery paper, as grit is very liable to work its way into the bearings.

In a seaway, the throttle requires a hand stationed to manipulate the valve so as to prevent violent "racing," unless an efficient governor is attached. Pumps and valves need attention after starting until they perform their functions as required. Steam and water gauge need constant inspection.

THUMPING.—Water in the cylinders, loose bearings, loose piston head, hot bearings, too little cushioning, slack cotters and pins, or loose valve gear, will be indicated by various noises which the engineer soon learns to distinguish.

LAYING UP.—Pick out the packing from stuffing boxes and plug up the oil holes and oil feeders with tallow or plugs, and cover up the bearings to keep out dirt. Clean down the engines and cover the polished parts with white lead and tallow. Turn or move the engines once in a while to prevent corrosion, and keep pipes, valves and elbows free from water.

X.

THE PRINCIPAL TYPES OF YACHT MACHINERY.

THE PERKINS HIGH PRESSURE SYSTEM.

ALTHOUGH the economy attained by the Perkins boiler and engine has of late been equalled in some cases, the system still stands at the head of modern practice so far as high pressure and extreme expansion of steam is concerned. The apparent failure of the Perkins system to reach such results as might be expected from steam at 300 and 400 lbs. pressure, is due to details in design of engines which interfere with the full realization in practice of the possibilities of the system. Many apparently insurmountable difficulties in the way of employing great pressures have been already successfully overcome, and if further improvements keep pace with the experience gained from the actual application of the system to marine purposes, there is reason to believe that the fuel consumption will be brought down to 1 lb. of coal per horse power per hour.

All objection to generating high pressures in a steam boiler have been met in the pipe arrangement illustrated in the chapter on Boilers. It combines a maximum of strength and safety, which can be supplied far beyond the critical necessities of each case. By the use of perfectly fresh or distilled water, incrustation and injury from acids is avoided, so that small tubes can be used without trouble or damage. These are arranged horizontally, with an internal diameter of 2¼ in. and 3 in outside, except the steam drum on top which is 4 in. internal and 5½ in. external diameter. The horizontal tubes are welded up at each end ½ in. thick and all are connected by

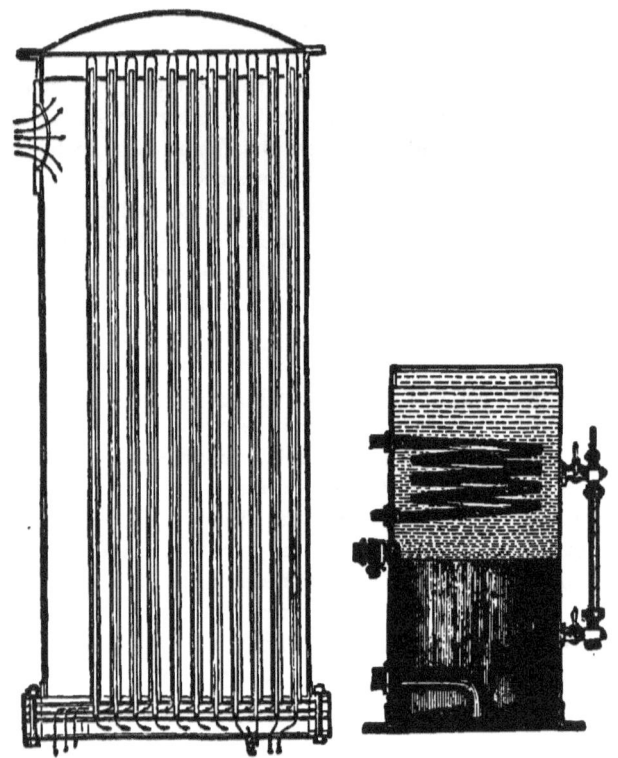

FIG. 57.—THE CONDENSER AND STILL.

small vertical joints of ⅞ in. internal and 1,¹⁶ in. external diameter. The fire box is built up of tubes in rectangular form, connected to one another by numerous vertical joints as shown in the cuts. The boiler is inclosed in a double casing of sheet iron, the space between being filled up with vegetable black or non-conducting substance.

Principal Types of Yacht Machinery. 157

Each tube is proved by hydraulic pressure to 4,000 lbs. and the boiler as a whole to 2,000 lbs. without showing leakage in several hours test.

The boiler can be worked with great regularity, without priming and with steam free from moisture. Repairs are easily made, for damage can only be local. If the feed gives out, the only chance of serious injury to the tubes can be prevented by dampening or drawing fires quickly as in any other boiler, while the disastrous results from explosion are reduced to a minimum from the small body of water and steam carried.

Distilled water is however a necessity to the preservation of the internal surfaces of the boiler. The feed water is used over and over again without admixture of sea water, by designing all parts of the machinery to be practically water-tight. The leakage is so small that 10 gallons per 100 Indicated H. P. per 24 hours will supply the waste. An actual test made upon engines indicating 250 H. P., with steam from the boiler at 250 lbs., developed no loss of water during thirteen days continuous running.

The exhaust steam from the engine is passed into a vertical surface condenser, the tubes of which are absolutely tight, the injection passing through the tubes and the exhaust steam surrounding them. The tubes are $\frac{7}{8}$ in. internal and $1\frac{1}{16}$ external diameter and welded up at the top end and securely fixed in a bottom tube plate. Internal tubes cause the water to circulate to the extreme ends as shown by the arrows in the cut. The small still is used to produce fresh water as a supply for such waste as may take place. Steam blowing off at the safety valve is returned to the condenser, and there may be a margin of 100 lbs. per sq. in. between the load on the safety valve and the pressure required to work the engines. The tubes of a boiler built in 1861 were cut in 1874 and submitted to official examination by the British Admiralty Boiler Committee. The tubes were in an excellent state of preservation after thirteen years use, with the longitudinal lines in them as sharp and well defined as in new ones.

The use of high pressures involved a remodeling of the engine as shown in the sectional diagram.

There are three cylinders, the first, A, is single-acting, and the second, B, likewise, having four times the capacity of the first, the pistons of both operating upon the same piston rod. The third cylinder, C, is double-acting, four times the capacity of the second, the piston rod being connected to a crank at right angles with the high-pressure crank. Lubricating the first cylinder and the use of glands and packing were out of question with such high pressure, hence the novel arrangement of the initial cylinder, whereby glands

FIG. 58.—SECTION OF PERKINS ENGINE THROUGH CYLINDERS.

are dispensed with. The steam is cut off at half stroke in A and exhausts under piston B for the up stroke. The temperature is by that time so much reduced that no serious difficulty is experienced in the use of the gland and packing. From the bottom of second cylinder B, the steam escapes into the top of the same, which serves as the intermediate chamber in connection with the valve chest of the third cylinder C. The latter is arranged to cut off at one-quarter stroke and exhausts into the condenser, the total expansion being $2 \times 4 \times 4 = 32$. All cylinders are steam-jacketed with coils, the condensed water of which is returned to the hot-well. The cylinders and valve boxes are also surrounded by a double sheet iron casing, filled with some non-conductor. The ordinary mode of packing the pistons and depending upon lubrication is superseded by the introduction of piston rings made of five parts of tin and fifteen of copper. Extensive practice in this and other engines has now established the perfect feasibility of the plan.

Thornycroft high-speed launches have been put through their trial trips of two hours duration, making 430 revolutions per minute, without the use of oil or grease, depending upon the rings of alloy only. The rings have been in use for nearly three years in stationary engines without lubrication of any sort. They have been a success on seagoing steamers running 10,000 miles, and showing cylinders smoothed up beautifully and giving no trouble thereafter. The steam yacht Anthracite made her transatlantic voyage without any trouble on the score of piston lubrication.

The Perkins engine and boiler is specially adapted to yachting purposes. The machinery is compact and takes up little space longitudinally. The boiler cannot burst with violence. It does not prime in a seaway and there is no noisy blowing off of steam. Engines can be stopped upon a moment's notice without risk of running up steam to dangerous pressure, as the tubes can be cooled to suit in a moment by attending the furnace door and damper. There is no smell and grease as no lubricants are used in the cylinders. The engine room can be kept as neat as the cabin.

Experiments made upon the steamer Anthracite by Mr. F. J. Bramwell, before her departure for America, gave an economic result of 1.83 lbs. of coal per I. H. P. per hour. The results obtained by dock trial at the Brooklyn Navy Yard were not as favorable as shown below, the consumption being 2.7 lbs. But this difference is fully acounted for in the official report made to the U. S. Navy Department by Chief Engineers Loring, Ayers and Magee, appointed to make the official test.

The difference of 36.88 per cent. in the two results is partly attributed to the use of inferior coal in the American trial, the English test having been made with standard Nixon· Navigation coal used in the measured mile runs of the British Admiralty, having only 5 per cent. of loose white ash against 17.6 per cent. of ash and clinker in the American Cumberland coal. Messrs. Loring, Ayers and Magee report that after allowing for this, there remained a difference of 17.64 per cent. superiority in Mr. Bramwell's trial to be accounted for, and this they attribute to their carrying the water at too high a level, losing the advantage of superheated steam. The piston speed was greater in the Bramwell experiment with less cylinder condensation in consequence, the difference being due to making the English trial under way and the American trial at the dock. It has also been stated that the first cylinder was not properly steam jacketed in the American trial, which seems probable from the fact that the condensation was found to be 34.99 per cent. by Mr. Bramwell against 56.22 by the U. S. Navy Committee. Calculations were made to discover whether the greater condensation in the American trial would represent the excess in fuel consumed and the committee report that such was the case. Furthermore, the engines were shown to be faulty in some details of construction, notably in the mode of steam jacketing and in the excessive clearances and steam passages, being from two to four times greater than required. This could be eliminated by a revision of the details of design, promising correspondingly greater economy in fuel consumption.

The Anthracite was built in 1878 by Schlesinger, Davis & Co., of Newcastle, and classed A1, 11 years, at Lloyds. She has an iron hull, with four bulkheads, registers 69.30 tons gross and 30 tons net. Her length on deck is 86 ft. 4 in., beam 16 ft., and depth of hold 10 ft. She left Erith on the Thames, England, May 29, 1880, and arrived at New York July 2, after touching at Falmouth and St. Johns, N. F. The passage from land to land was made in 18 days, the whole distance under steam only, through heavy weather and rough sea. The engines have already been described. She used 436 gallons of distilled water to supply the waste during her trip across. Cylinders were $7\tfrac{3}{4}$, $15\tfrac{13}{16}$ and $22\tfrac{13}{16}$ in. bore, with 15 in. stroke. Diameter of piston rods to be deducted from the areas of second and third cylinders, is $2\tfrac{3}{4}$ in.

The valve motion is derived from eccentrics, the low-pressure cylinder having an ordinary slide valve with an expansion valve on

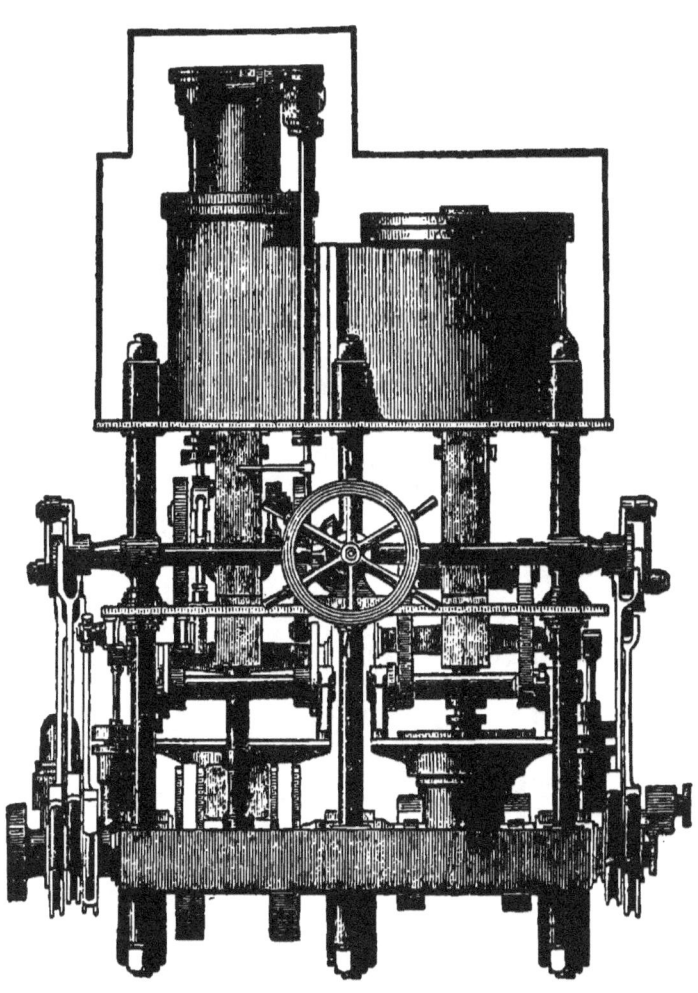

Fig. 59.—Engines of S. S. Anthracite. Front Elevation.

FIG. 60.—ENGINES OF S. S. ANTHRACITE. SIDE ELEVATION.

its back worked by the prolongation upward of the circulating pump rod. Steam is cut off in the high-pressure cylinder by three lifting double-beat valves, the upper faces being divided into sections. The surface condenser is supplied with galvanized wrought iron tubes of 422 sq. ft. surface.

The circulating pump is 11½ in. diameter, worked by a beam off the low-pressure piston rod, the air pump, of like diameter, being actuated by a similar beam off the intermediate cylinder piston rod. The two feed pumps, of 2 in. diameter, and the two bilge pumps, of 3 in. diameter, are actuated from the crossheads of the circulating and air pumps. The engine is reversed by screw hand wheel and link gear.

The boiler has 15 sq. ft. of grate area and 633 sq. ft. of heating surface, the pressure ranging from 300 to 500 lbs.

From the U. S. Government report of the Anthracite's trial, tied up to the dock at the Brooklyn Navy Yard, the following data are obtained, consumption of coal having already been considered:

Pounds of coal consumed per hour per indicated horse power........... 2.7115
Total quantity of Cumberland semi-bituminous coal consumed, in pounds 4,400
Total pounds of refuse in ash and clinker........................ 776
Total pounds of combustible consumed............................. 3,624
Total pounds of feed water pumped into the boiler.................... 35,114
Total double strokes of the pistons................................. 148,154
Steam pressure in the boiler, in pounds, above the atmosphere 316.50
Steam pressure in the cylinder, in pounds, above the atmosphere 210.54
Throttle wide open
   (In none of the cylinders was the steam cushioned, nor was there
   steam or exhaust lead.)
Vacuum in condenser, in inches of mercury.......................... 26.75
Back pressure in condenser, in pounds, above zero................... 1.6066
Temperature, in degrees Fahr., of feed water........................120.5
Temperature, in degrees, of steam in the boiler, saturated............420.0
Pounds of coal consumed per hour..................................183.5883
Pounds of coal consumed per hour per square foot of grate............ 11.9867
Mean back pressure against the piston of third cylinder, in pounds...... 4.21
Indicated horse power of 1st cylinder............................... 20.4308
   "   "   2d   " 7.8290
   "   "   3d   " 39.4483
Aggregate indicated horse power for all three cylinders............... 67.7081
Total horse power developed in three cylinders....................... 80.1525

Pounds of feed water consumed per hour per indicated horse power..... 21.63875
Pounds of steam condensed in the first, second and third cylinders, to furnish the heat transmitted into the total horse power developed in those cylinders, by the expanded steam alone 167.0720
Pounds of water vaporized from 212° by one pound of coal 9.2671
Duration of trial in hours..,........ 24
Total weight of machinery, engines, boiler, screw shaft, propellor and all fittings, in tons 25

THE HERRESHOFF SYSTEM.

The coil boiler, upon which the Herreshoff system rests, has all the advantages of the pipe arrangement described in the Perkins system, and in addition the very material one of light weight in comparison to the power. This makes the coil especially suited to high speed yachts, as the saving in weight enables refinement of form in hull beyond the displacement which is a necessity for heavier steam generators. The coil boiler is also particularly well adapted for launches and cutters, owing to the quick raising of steam and the possibility of hoisting such boats to the davits of a large steam yacht, without removing the machinery and without undue strain. An extensive report concerning the Herreshoff method was made to the Bureau of Steam Engineering, U. S. Navy Department, in 1881. The experiments were carried out upon the steam yacht Leila by Chief Engineers B. F. Isherwood, Theo. Zeller and Geo. W. Magee. It is worth noting that this was done at the invitation of the Herreshoff Manufacturing Co., of Bristol, R. I., and that they freely undertook the entire expense attendant upon the trials. These have since been frequently referred to by the engineering world in making comparisons with other machinery, and the liberality of the manufacturers stands as an example which other builders of special systems might follow with benefit to all concerned. From the official report much of the following description is extracted. It should be added that their practice in building compound engines has been of late supplemented by new patterns of triple expansion engines, for which the Herreshoff system is equally well adapted.

The steam yacht Leila, as an example, is of composite construction, the frame being of angle iron planked with Southern pine and

Principal Types of Yacht Machinery. 165

sheathed with copper; the stem and sternpost are of wood. The waterlines of the hull are excessively sharp and the angle of its dead rise proportionally great.

The draft of water of the hull proper at the stern was so small that the screw had partly to descend below the bottom of the keel in order to be wholly immersed; this required the addition of a skeg at the stern and below the keel, for the purpose of protecting the screw and of supporting the metallic shoe which extends horizontally beneath it. The aftermost end of this shoe sustains the lower pintle of the rudder; the latter is of copper and counterbalanced with the axis at one-fourth of its breadth from the forward edge.

The skeg is a right-angled triangle of wood, 15½ in. deep below the bottom of the keel, at the after side of the sternpost, and 60 in. long upon the bottom of the keel from the after side of the sternpost; its breadth is 7 in.

The following are the principal dimensions and proportions of the hull:

Extreme length on top of deck .... 100 ft.
Length on waterline from forward edge of stem to after side of sternpost .... 95 ft. 5 in.
Extreme breadth on top of deck .... 15 ft. 4 in.
Extreme breadth on waterline .... 11 ft. 9 in.
Depth of hull amidship from lower edge of rabbet of keel to top of
  deck beams .... 5 ft. 10 in.
Depth at stern from waterline to lower edge of rabbet of keel .... 3 ft. 1½ in.
Depth of keel aft, below lower edge of its rabbet .... 8½ in.
Siding of keel .... 7 in.
Distance from the forward edge of the stem to the greatest immersed
  transverse section .... 54 ft. 6 in.
Area of the waterline .... 699.5 sq. ft.
Area of the greatest immersed transverse section, including projected
  areas of keel and skeg .... 20.12 sq. ft.
Displacement (35 cu. ft. per ton) .... 37.27 tons.
Aggregate area of wetted surfaces of hull, keel and skeg .... 935.5 sq. ft.
Angle of dead rise at greatest immersed transverse section .... 21½ deg.
Half angle of bow on waterline .... 8½ deg.
Half angle of stern on waterline .... 14 deg.
Ratio of the length to the breadth on waterline 8.1206
Ratio of waterline plane to circumscribing parallelogram 0.6239
Ratio of the greatest immersed transverse section to its circumscribing
  parallelogram 0.5624

Ratio of the displacement above lower edge of rabbet of keel to its circumscribing parallelopipedon............................0.3721
Distance between centers of angle-iron frames.....................1 ft. 3 in.
Angle-iron frames, molded.......................................1½ in.
Angle-iron frames, sided..2¼ in.
Thickness of iron of angle-iron frames.......................... ¼ in.
Deck beams, molded ....2 in.
Deck beams, sided ....3½ in.
Thickness of wooden stem ....6 in.
Thickness of wooden sternpost ....7 in.
Thickness of bottom and side planking...........................1¼ in.
Thickness of deck planking ....1 3/16 in.

The Leila has one boiler with single circular furnace, around and over which the continuous pipe of wrought iron composing the heating surface is coiled spirally and symmetrically into two concentric coils, one coil being immediately on the outside of the other so as to surround it. This pipe contains the water to be evaporated, and the hot gases of combustion act on its exterior, enveloping every part of it from one end to the other. The grate, circular in plan, is inclosed by a circular wall of brick masonry, on the top of which the double coil rests, and the latter is surrounded by two concentric casings of sheet-iron, with an air space between. The whole of the gases of combustion pass between the spirals from the inside to the outside of the two coils into the space between the latter and the casing, and from this space these gases ascend the chimney. The feed water enters the inside coil at the extreme upper end, whence it flows partly by gravity, but mainly by the action of the feed pump, down to the extreme lower end of this coil; thence into the extreme lower end of the outside coil, up which it ascends to the extreme upper end, being converted in its progress into steam. If the supply of feed water relatively to the heat of the furnace be such that the former is entirely vaporized by only a portion of the heating surface in the coils, then the remaining portion of that surface will act to superheat the steam. As the latter effect should be avoided on account of the injurious action of the intense direct heat of the furnace on the iron of the pipe when unprotected by water, recourse is had to a forced circulation of a superfluous quantity of feed water by means of a circulating pump, which, by continually drawing this superfluous feed water from the delivering end of the double coil and forcing it into the receiving end, keeps both coils always sufficiently filled with water to prevent steam superheating, let the quantity of water vaporized be what it may. The feed and the superfluous feed both enter simultaneously and at the same point. The mixed water and steam are projected from the delivering end of the coil pipe into the "separator," which is merely a closed cylindrical vessel where the water, by its greater gravity, separates from the steam and falls to the bottom, while the steam is carried off from the top by a pipe, which, after winding spirally two and five-eighths times around the upper portion of the inside coil, appearing like an extension of the upper portion of the outside coil on which it rests, passes to the engine. All the surface in these two and five-eighths spirals is steam-superheating surface, which, being exposed to the gases of combustion at high temperature, acts very efficiently for that purpose. The water collected in the "separator" is again pumped by the circulating pump into the top of the pipe of the inside coil. The "separator" acts both as a vessel in which the separation of the water and steam takes place and a steam drum for the storage of steam at approximately constant pressure. In this boiler the water and steam occupy exactly opposite positions to what they do in ordinary boilers, the water being in the top and the steam in the bottom of the coil.

Upon the "separator" are placed the safety valve, the steam-pressure gauge, and a glass water gauge for showing the height of the water in the lower portion of the "separator." This height is the water level to be carried, and its maintenance regulates the quantity of superfluous feed water to be pumped in by the circulating pump. By properly proportioning that pump, any amount of superfluous feed water can be kept in circulation, and the current forced over the heating surfaces in such a torrent as to sweep off the steam bubbles as fast as formed, and to change and mix the water with such rapidity as to obtain the maximum heating efficiency from a given area of those surfaces in a given time. The glass water gauge on the "separator" answers the same purpose as the gauge cocks on boilers of the usual construction, and requires to be as closely watched, for on the continuous passage through the coil pipe of an excess of feed water over what is vaporized depends the preservation of the metal from burning.

The furnace consists of a circular grate 5 ft. and 9 in. in diameter, surrounded by a circular vertical wall of fire brick laid in fire clay. The opening for the ashpit door is 13 in. wide and 11 in. high. Opposite the door is another opening for receiving the blast from a fan blower.

Upon the top of the brick wall inclosing the furnace rests the double coil of continuous wrought iron pipe. The inner coil may be conceived to be wound spirally around the frusta of two right cones, one super-imposed upon the other, and having their axes in the same vertical line. These imaginary frusta form the inner dimensions of the inside coil. The lower one is 5 ft. 5 in. in diameter at bottom, 4 ft. 8 in. in diameter at top, and 4 ft. 3 in. in height. The upper one is 4 ft. 8 in. in diameter at bottom, 1 ft. 4 in. in diameter at top, and 8 in. in height. Above these two frusta the pipe forming the inside coil is extended into a nearly horizontal spiral of 6 ft. 10 in. in extreme diameter, formed of eleven and a half circumvolutions, and placed as low as the upper frustum will allow. This horizontal spiral, situated in the uptake and above the coils, is exposed to the hot gases of combustion just before they enter the chimney, and after as much of their heat as possible has been extracted by the coils with which they first come in contact. Consequently, it acts as a heater, the feed water and circulating water being delivered into it at one extremity, forced round the spirals by the feed and circulating pumps, and emerging from it into the inside coil at the other extremity, with which it is continuous.

The pipe composing the horizontal spiral is of wrought-iron, lap-welded, 1⅞ in. outside diameter and 1½ in. inside diameter; thickness of metal $\frac{3}{16}$ in. The exterior surface of the horizonal spiral is 73.63 sq. ft., its interior surface is 58.90 sq. ft., and its content is 1.84 cu. ft.

The inside coil, starting from the top, is composed of four turns or helical spirals of wrought-iron lap-welded pipe, 1⅞ in. outside diameter and 1½ in. inside diameter, thickness of metal, $\frac{3}{16}$ in.; five

spirals of 2⅜ in. outside diameter and 2 in. inside diameter, thickness of metal, 1/16 in ; eight and a half spirals of 2⅞ in. outside diameter and 2½ in. inside diameter, thickness of metal 3/16 in.; and five and a half spirals of 3½ in. outside diameter and 3 in. inside diameter, thickness of metal ¼ in. The four spirals of 1⅞ in. outside diameter pipe and the five spirals of 2⅜ in. outside diameter pipe form the top of the furnace and are in close contact, so that none of the gases of combustion can pass between them. The eight and a half spirals of 2⅞ in. outside diameter pipe and the five and a half spirals of 3½ in. outside diameter pipe form the sides of the furnace, and are separated by spaces through which all the gases of combustion pass. The exterior surface of the inside coil is 239.12 sq. ft., the interior surface is 204.98 sq. ft., and the contents 10.93 cu. ft.

The outside coil, continuous with the inside coil and connecting with it at the bottom, is composed of nine and three-fourths spirals of wrought-iron lap-welded pipe 3½ in. outside diameter and 3 in. inside diameter, thickness of metal ¼ in. The space between the outside coil and the inside coil is 1½ in. The exterior surface of this coil is 172.26 sq. ft., the interior surface is 147.65 sq. ft., and the contents 9.23 cu. ft.

Total water heating surface of the boiler is 485.02 sq. ft. exterior pipe surface, with 411.56 sq. ft. interior surface, and contents of 22.00 cu. ft.

The superheater on top of outside coil has 43.98 sq. ft. exterior surface, 37.70 sq. ft. interior surface, and 2.36 cu. ft. contents.

All the gases of combustion, after passing between the spirals of the two coils and the superheater, impinge on the inside casing, which has thus nearly their temperature.

The uptake rests upon the casings, and is composed like them of two parallel sheet iron plates 1/16 in. thick, with a ⅝ in. intervening space filled with mineral wool. The form of the uptake is a frustum of a right cone, 7 ft. in diameter at bottom, 2 ft. in diameter at top, and 7 in. in height.

The chimney rests upon the uptake, and is 2 ft. in diameter and 20 ft. in height above it.

Principal Types of Yacht Machinery. 171

The "separator" is placed by the side of the boiler, with a space of 7½ in. between them. It is simply a hollow cylinder formed of ⅜ in. thick boiler plate, and has both top and bottom closed. It receives at the top the mingled water and steam from the top or delivering end of the outside coil, which is prolonged into the "separator" for about one-third the height of the latter. In the "separator" the water is separated from the steam by gravity and falls to the bottom, while the steam is carried off from the top of the "separator" by the superheating coils which lead it to the engine. The "separator" is thus an independent but essential adjunct of the boiler, intended to act both as a steam reservoir or steam drum and as a water trap. The top is fitted with a safety valve of the usual construction, and the bottom is fitted with a blow-off pipe and cock for draining the "separator" and blowing out any sediment that may collect in it. On the side of the lower portion of the "separator" is a glass water gauge of the usual construction, which shows on inspection whether there is an excess of feed water passing through the coils of the boiler; the proper performance of the boiler requiring always such excess. From near the bottom of the "separator" a pipe proceeds to the circulating pump, which continually removes this excess of feed water and forces it back into the boiler by a pipe connecting the delivery of the pump with the receiving end of the heater coil.

The principal dimensions of the Leila's boiler are summed up below:

Diameter of the boiler to outside of casing....................84 in.
Height of boiler from bottom of ashpit to top of uptake.........99 in.
Diameter of the furnace.......................................69 in.
Area of the grate surface.................................... 25.9673 sq. ft.
Area of water-heating surface measured on outside of coil pipe...485.01723 sq. ft.
Area of steam superheating surface measured on outside of coil
    pipe... 43.9824 sq. ft.
Cross area of chimney.. 3.1416 sq. ft.
Height of the chimney above the level of the grate bars.........27 ft.
Interior diameter of the "separator '.........................14 in.
Interior height of the "separator ....50 in.
Total steam room in separator and superheater................. 6.0083 cu. ft.
Aggregate contents (water and steam) of the heater and of the
    inside and outside coils..................................21.99793 cu. ft.

Square feet of water-heating surface, measured on outside of coil pipe, per square foot of grate surface.......................18.6780
Square feet of steam-superheating surface, measured on outside of coil pipe, per square foot of grate surface................... 1.6938
Square feet of grate surface per square foot of cross area of chimney for the passage of the gases of combustion............. 8.2656

The engines of the Leila are vertical compound condensing, the cylinders being direct acting. The small cylinder operates a lever which works the air pump, the feed pump, and the circulating pump, all of which are vertical, single acting, and have the same stroke of piston. The air pump discharges into an open-topped hot-well or reservoir placed above the outboard waterline, the top of the air pump being closed.

The cylinders are separated to allow the valve chests to be placed between them, with sufficient additional space for the removal of their covers. The valves of each cylinder are a plain three-ported slide with a slide cut-off on its back; these valves are not counterbalanced, but work with the full steam pressure on their backs. The three-ported slides or steam valves are operated each by a Stephenson link and two eccentrics, which serve as a reversing gear. The cut-off valves are each operated by an eccentric. The cut-off valve of the small cylinder is adjustable; that of the large cylinder is fixed to cut off at about one-third of the stroke of the piston from the commencement.

The engine works with surface condensation. The surface condenser is composed of a single copper pipe placed on the outside of the vessel, beneath the water, and just about at the garboard strake. This pipe commences on one side of the vessel abreast of the after or large cylinder, extends to and around the sternpost, and thence along the opposite side of the vessel until abreast of the air pump and forward cylinder. The diameter of the pipe continuously decreases from the end at which it receives the exhaust steam from the large cylinder to the end at which it delivers the water of condensation and the uncondensed vapor and air into the air pump whence they are thrown into the hot-well from which the feed pump forces the water of condensation into the top of the boiler coil, where it is revaporized, and the steam, passing first into the small cylinder and thence into the large one, is finally exhausted into the condensing tube. It is essential for satisfactory working that the delivering end of this tube should not exceed one-half the diameter of its receiving end; for if a larger diameter be given to the delivering end, a part of the exhaust steam will pass directly to the air pump over the water of condensation in the tube. The delivering end of the tube must be small enough to remain completely filled with water for the exclusion of the steam from the pump.

The after main pillow-block serves also as the thrust pillow-block, the after journal of the crank-shaft being made with the necessary thrust rings upon it.

From the "separator" the circulating pump continuously draws water and forces it into the top of the boiler coil, where it enters along with the feed water from the feed pump, thus maintaining a forced circulation through the coil of what may be termed "superfluous feed."

Details of the engine:

Diameter of the small cylinder................................9 in.
Diameter of the piston rod of the small cylinder...............1⅝ in.
Net area of the piston of the small cylinder...................62.58 sq. in.
Stroke of the piston of the small cylinder.....................18 in.
Length of steam port of small cylinder.........................7.5 in.
Breadth of steam port of small cylinder........................1 1/8 in.
Area of steam port of small cylinder ....7.97 sq. in.
Length of exhaust port of small cylinder.......................7.5 in.
Breadth of exhaust port of small cylinder......................1.75 in.
Area of exhaust port of small cylinder.........................13.125 sq. in.
Clearance of piston of small cylinder..........................7/8 in.
Diameter of the large cylinder.................................16 in.
Diameter of the piston rod of the large cylinder...............1⅝ in.
Net area of the piston of the large cylinder...................200.02545 sq. in.
Stroke of the piston of the large cylinder.....................18 in.
Length of steam port of large cylinder.........................13 in.
Breadth of steam port of large cylinder........................1 7/16 in.
Area of steam port of large cylinder...........................18.69 sq. in.
Length of exhaust port of large cylinder.......................13 in.
Breadth of exhaust port of large cylinder......................2.5 in.
Area of exhaust port of large cylinder.........................32.5 sq. in.

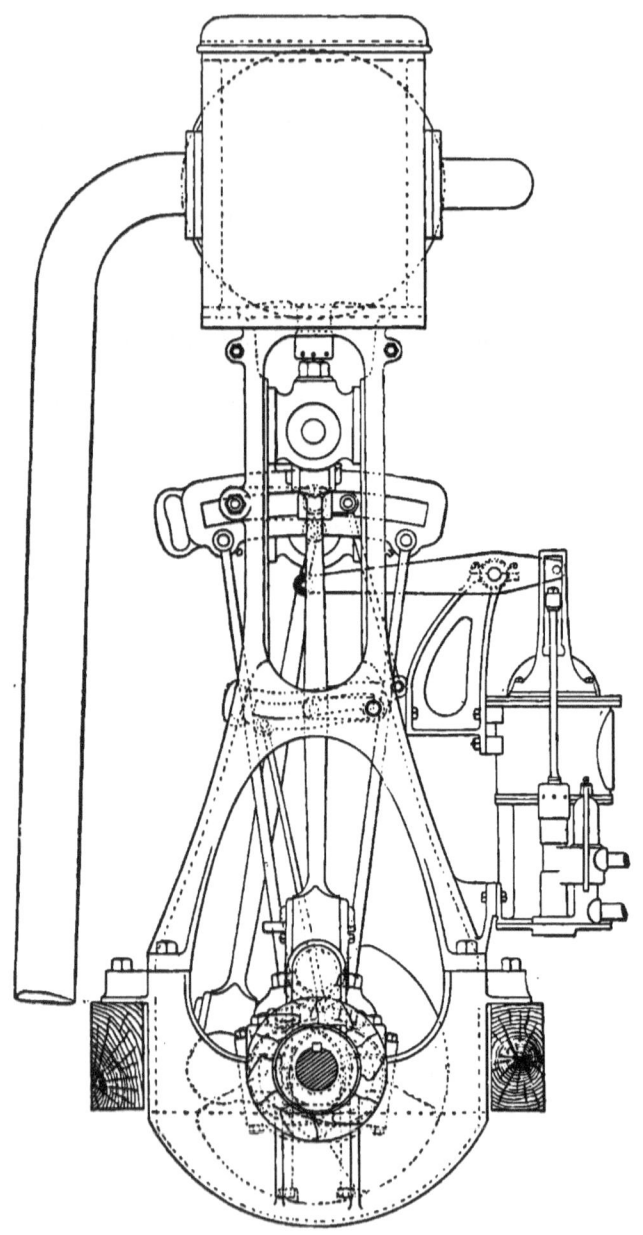

Fig 62.—Herreshoff Engine—End Elevation.

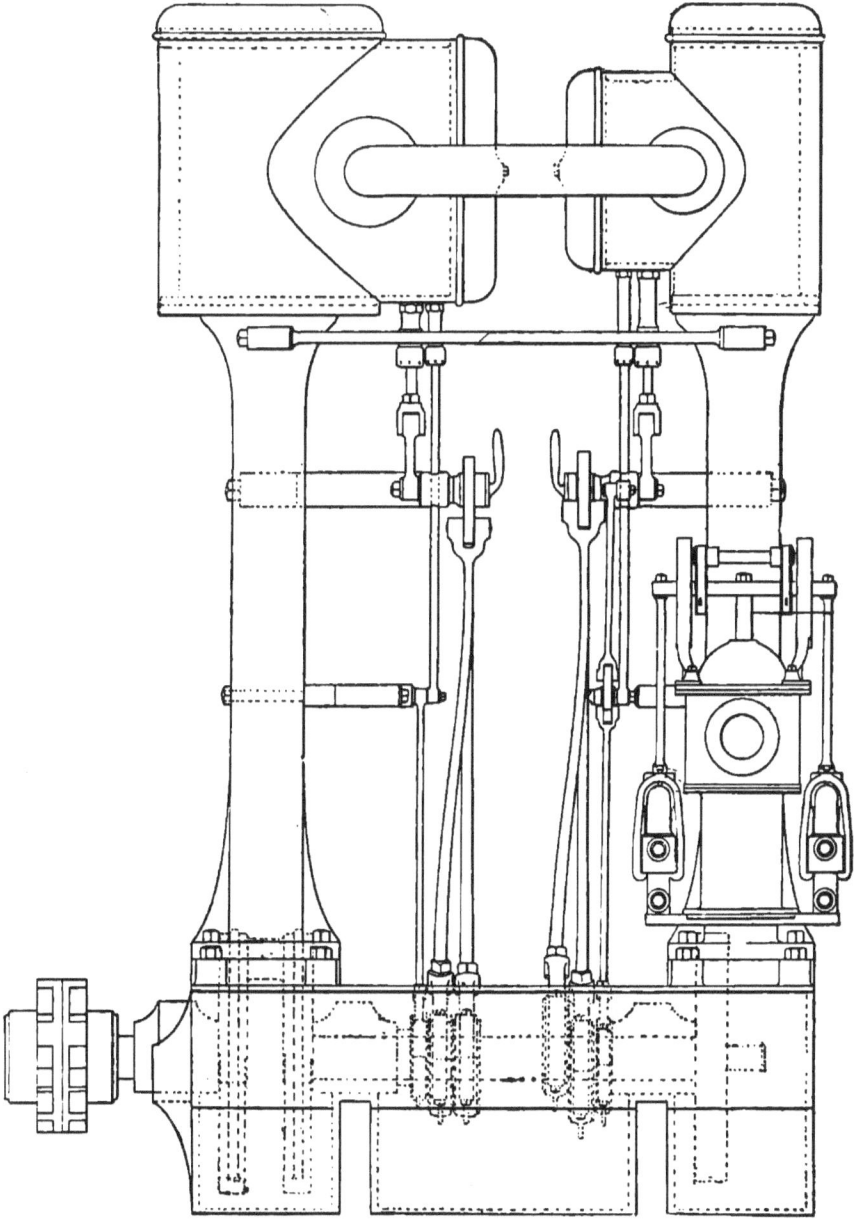

FIG. 63.—HERRESHOFF ENGINE—SIDE ELEVATION.

Clearance of piston of large cylinder.........................1/16 in.
Diameter of the air pump.....................................7 in.
Stroke of the piston of the air pump.........................6 in.
Diameter of the plunger of the feed pump.....................1½ in.
Stroke of the plunger of the feed pump.......................6 in.
Diameter of the plunger of the circulating pump..............1⅛ in.
Stroke of the plunger of the circulating pump................6 in.
Length of the condensing pipe ....53 ft.
Inside diameter of condensing pipe at exhaust steam end......5 in.
Continuous decreasing to inside diameter at air pump end of..2 in.
Thickness of the metal of the condensing pipe (copper).......1/16 in.
Exterior surface of the condensing pipe......................50.2983 sq. ft.
Length of the connecting rods between centers................49½ in.
Diameter of screw-shaft inside of brass casing...............3⅞ in.
Length in the vessel occupied by the engine..................66 in.
Breadth in the vessel occupied by the engine.................48 in.
Height of engine above center of crank shaft.................96 in.

The screw of the Leila is of brass and four-bladed, the pitch being uniform, the leading and following edges are parallel, when projected on a plane parallel to the axis. The diameter is 4 ft. 7 in.; diameter of hub 7 in. Pitch 8 ft.; projected area upon a plane at right angles to the axis is 6.59 sq. ft.

A summary of the Leila's performance is given below. The vessel was not forced to highest speed, but worked at ordinary rate under natural draft only. Artificial draft would have given higher speed of vessel at increased expenditure in fuel. Steam was cut off at 0.4 of stroke in small cylinder and 0.36 in large cylinder.

Steam pressure in boiler.....................................129 lbs.
Number of times steam was expanded...........................7
Vacuum in air pump...25.8
Number of revolutions per minute.............................221.5
Piston speed...664.5
Speed of vessel per hour in miles............................15.50
Speed of vessel per hour in knots............................13.45
Slip of screw in per cent. of speed..........................23
Initial pressure in small cylinder...........................135 lbs.
Initial pressure in large cylinder...........................45 lbs.
Indicated horse power..150
I. H. P. per ton of displacement.............................4.02
Weight of boiler and contents per I. H. P....................72 lbs.
Heating surface per I. H. P..................................3.5 sq. ft.

Principal Types of Yacht Machinery. 177

Grate surface per I. H. P..................................0.173 sq. ft.
Coal consumed per I. H. P. per hour............................2.21 lbs.
Coal consumed per square foot of grate..........................12.8 lbs.
Coal consumed per square foot of heating surface..................0.68 lbs.

The coal used was common anthracite, leaving 15 per cent. of clinker. With selected coal of the same grade as Nixon's Navigation, the consumption would have been reduced to 2 li's. per I. H. P. per hour. The maximum speed recorded for yachts of the Leila class is 22.8 miles or 19 knots.

This has since been surpassed in the latest yachts of Herreshoff construction, the most notable of which are the launch Henrietta and the Stiletto, now the property of the U. S. Government.

The Stiletto is very light but strong, a frame of bent oak well fastened to heavy keel, and garboards with two heavy oak wales on each side, making a very strong framework, which is covered with a double skin of white pine, with decks of the same material. The hull is divided by watertight bulkheads. The most peculiar feature is the upper portion above the hull proper. Instead of the ordinary deck and cabin trunk the sides are carried up, as shown, beveling slightly, high amidships and low at the ends, the crown being almost the reverse of the sheer line. These two sides each form a trussed girder, stiffening the entire hull, while the weight is mostly amidship. From their shape they offer little resistance to the wind. Her dimensions are: Length over all, 94 ft.; beam, 11 ft., (a proportion of $8\frac{6}{11}$), and depth of hold, 7 ft. 9 in.

The engine is compound, 12 and 12 × 12, capable of 450 turns per minute. Annular valves are used, cutting off generally at ⅝. The weight of the engines is 4,275 lbs., and they can work up to 450 H. P. with forced draft. Many of the parts are of steel, the shaft being 4 in. diameter. The wheel is 4 ft. diameter with 6 ft. 6 in. pitch, four-bladed. The boiler is a sheet-iron box 7 ft. square, set on a fire-brick foundation; the upper part of this box tapers into the stack, like an inverted mill hopper. Inside the fire box is 6 ft. 3 in. square. Just above the fire is a row of tubes 3½ in. in diameter, running side by side thwartship, each tube being con-

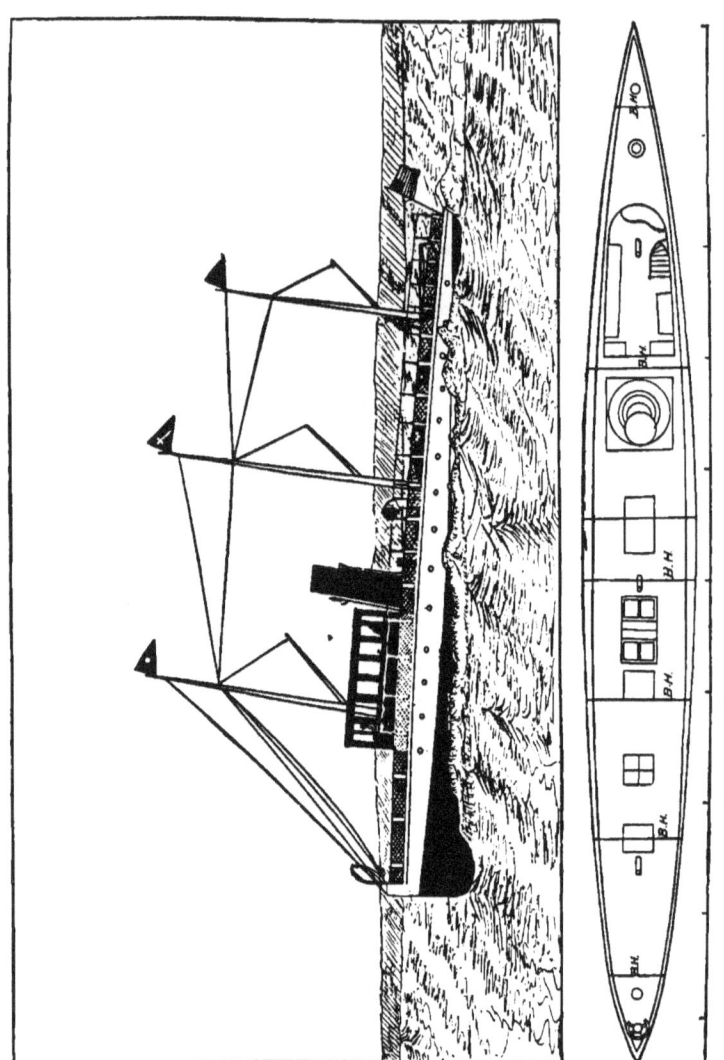

Fig. 64.—Herreshoff High-Speed Steamer Stiletto.

nected to its neighbor at alternate ends. Above are six other sets, decreasing in size to 1½ in. diameter, the second set running fore and aft, the third parallel with the first, etc., making practically one long tube, folded compactly into a small space. The upper or smaller end of this coil is connected with the feed pump, and the lower and larger end with the separator, a vertical cylinder 4 ft. long and 18 in. diameter, placed in front of the boiler. On this are the gauges and cocks. The water and steam entering here from the pipes are separated, the latter passing to the high-pressure cylinder. The water is used continuously, the only waste being by the whistle or leakage. The water pumped into the boiler at the top of the coil, converted into steam as it descends through the pipes, passes to the separator, thence to the high-pressure cylinder, the receiver, low-pressure cylinder, and condenser, and finally to the feed pump and boiler, any loss being supplied by an injector. The heating surface is 615 sq. ft. and the boiler works up to 160 lbs. pressure. The total weight is 13,637 lbs. The consumption of coal is about 2 lbs. per H. P. per hour. The displacement of Stiletto is 28 tons.

VERTICAL DIRECT-ACTING ENGINES.

In direct-acting simple engines, vertical tubular or horizontal tubular boilers, the long experience of the New York Safety Steam Power Company has placed their yacht machinery in the front rank in the East, and many examples of their work can be met with in foreign waters. The engines, of which an illustration has been given in the first chapter, are designed and constructed to possess great strength, with compactness of form, freedom of action, simplicity and perfection of details. Accurate workmanship and fine finish play an important part in performance and wear, particularly at the high speeds now requisite, when journals are apt to heat and breaking down is not infrequent. In grace of form and proper distribution of material the engines under consideration are without rival. For launches and yachts not intended primarily for distant cruising, the absence of the compound feature is not without its

FIG. 65.—STEAM LAUNCH KATRINA. BUILT BY N. Y. SAFETY STEAM POWER CO.

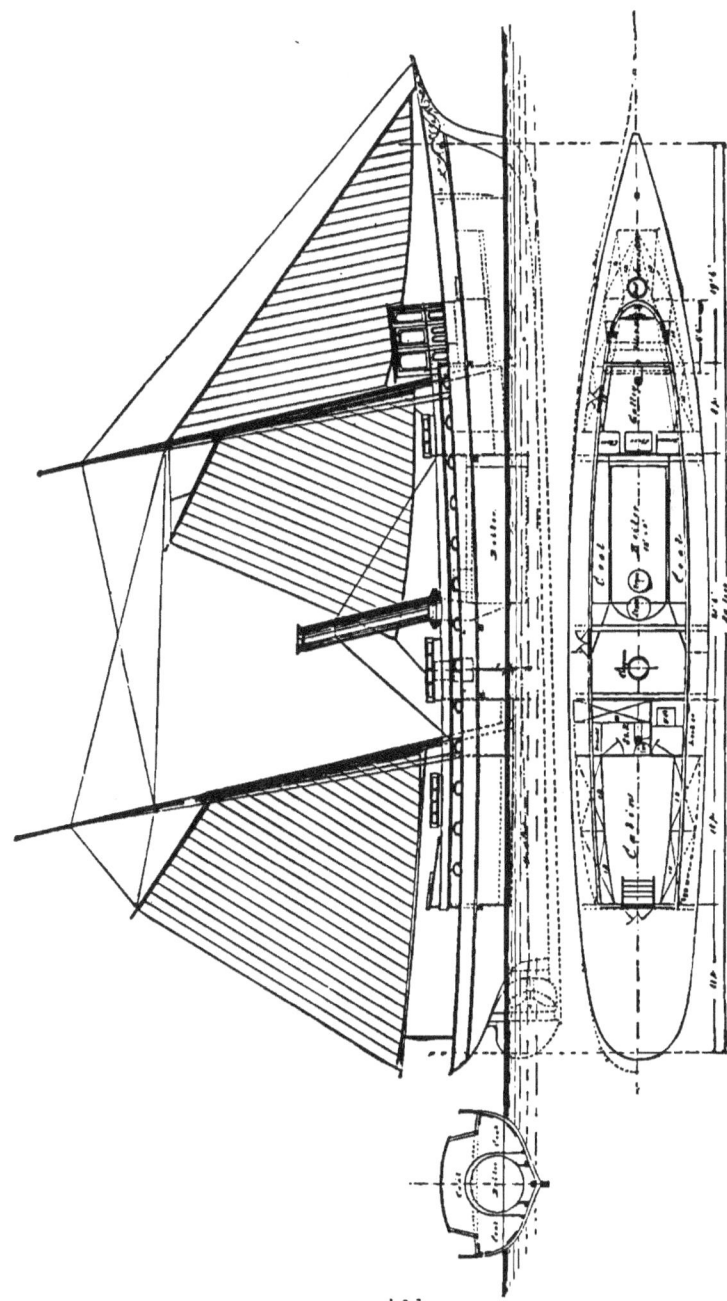

Fig. 66.—Steam Yacht. N. Y. Safety Steam Power Co.

Length on waterline, 72 ft. Beam, extreme, 13 ft. Depth of hold, 5 ft. 9 in. Forward cabins, 20 ft.
Length on deck, 85 ft. Beam on waterline, 11 ft. 9 in. Draft aft, 5 ft. 3 in. After cabins, 18 ft.
Boiler, length, 12 ft. Engine, 10x12 in. Machinery space, 21 ft.
Boiler, diameter, 5 ft.

advantages. Fuel economy is not such an important item in such boats as reliability and moderate first cost.

The vertical engine has two bearings for the crank shaft in the frame, both being cast solid with the standard or column, as are also the cross-head slides. There can be no derangement of the line. On larger sizes an adjustable thrust-bearing is secured to the engine frame in addition. The cranks are counterbalanced, crank shaft, piston rod, valve stem, cross head, pin, etc., are of steel. The link motion has a cut-off index, by which the expansion can be regulated. In engines larger than 7×9 the main valve is "balanced," that is, so arranged that it is not subject to the steam pressure on its back in the steam chest and therefore not pressed hard against the valve seat, thus obviating undue friction. The steam and exhaust openings are double, so that the steam may have free exit and entrance and preserve the initial pressure in cylinder near that of the boiler. Engines are fitted with link reversing gear, cylinder lubricator, stop valve, drip cocks, pry wheel for starting when on the dead center, couplings and oil cups. The details of some leading sizes are given in the table:

Size of Cylinder Diameter. Stroke.	Diameter of Crank Shaft.	Height from Floor to Top of Cylinder	Weight of Engine.	Diameter of Base.	Size of Boat for Which Suitable. (Approximate.)		
					Length.	Beam.	Draft.
Inches.	Inches.	Inches.	Pounds.	Inches.			
3 × 5	1¾	36½	230	13	25 ft.	5 ft. 8 in.	27 in.
3½× 5	1¾	36½	270	13	28 ft.	5 ft. 10 in.	28 in.
4 × 6	2	44½	295	15½	30 ft.	6 ft.	30 in.
5 × 6	2	44½	315	15½	32 ft.	6½ ft.	32 in.
5½× 7	2½	51	460	18	35 ft.	7 ft.	35 in.
6½× 7	2¾	51	545	18	38 ft.	7½ ft.	38 in.
7 × 9	3	63½	825	26	40 ft.	8 ft.	40 in.
8 × 9	3	63½	1200	26	45 ft.	10 ft.	48 in.
9 ×12	4	77½	2200	26	65 ft.	11 ft.	54 in.
10 ×12	4	77½	2800	26	70 ft.	12 ft.	60 in.
12 ×12	5	90	3500	32	75 ft.	15 ft.	60 in.

The applicability of the screw to extreme light draft is illustrated in the plans of a launch 32 ft. long and 8 ft. beam. To obtain enough area of wheel upon only 12 in. immersion, three wheels take the place of the usual single or even twin screws, all three being driven by gears from the single-cylinder engine. All the wheels are kept above the keel line, and ño skag is introduced. The launch may therefore "take the ground" without risk, an immunity which will be appreciated by sportsmen or others navigating unknown waters or where shifting bars render charts unserviceable.

FIG. 67.—OPEN STEAM LAUNCH.

Launch boilers are of the vertical tubular style, being compact, light and simple, and also as economical and safe as any other style when built to good proportions. For decked yachts the horizontal fire tubular boiler is preferred, owing to its lower center of gravity. The mountings and attachments comprise ash pan, grates, smoke bonnet, stack with guys, safety valve, gauges, blow-off cocks, feed check, feed globe valve, stop valve and whistle. The hydraulic test is 200 lbs. Proportions of vertical boilers are given in the table:

Diameter of Boiler.	Height of Boiler.	Tubes.		Heating Surface, Square Feet.	Size of Engine Cylinder to be Driven.
		Number of.	Diameter.		
28	48	84	$1\frac{1}{4}$	60	3×5
30	48	150	$1\frac{1}{4}$	95	$3\frac{1}{2} \times 5$
33	48	180	$1\frac{1}{4}$	104	4×6
36	48	204	$1\frac{1}{4}$	137	5×6
36	$54\frac{1}{2}$	204	$1\frac{1}{4}$	197	$5\frac{1}{2} \times 7$
44	66	120	2	205	$6\frac{1}{2} \times 7$
46	76	136	2	256	7×9
48	82	144	2	280	8×9
50	82	180	2	380	9×12
54	85	204	2	430	10×12

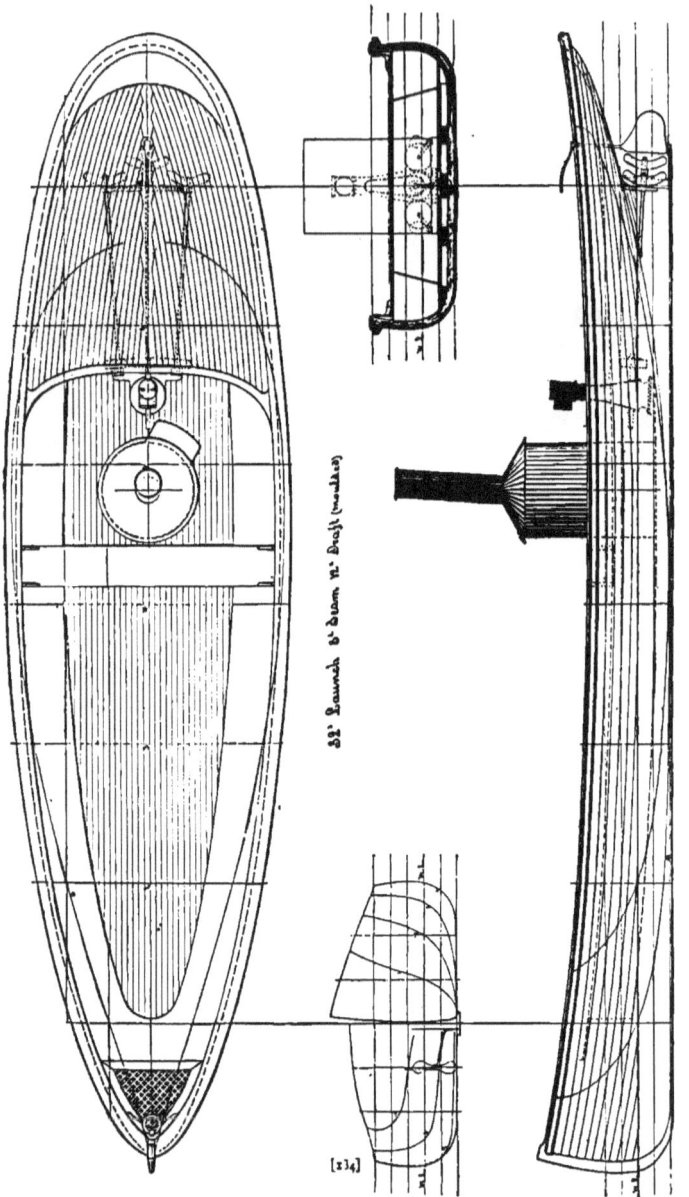

FIG. 68.—TRIPLE SCREW LAUNCH, EXTREME LIGHT DRAFT.
Length, 32 ft. Beam, 8 ft. Draft, 12 in.

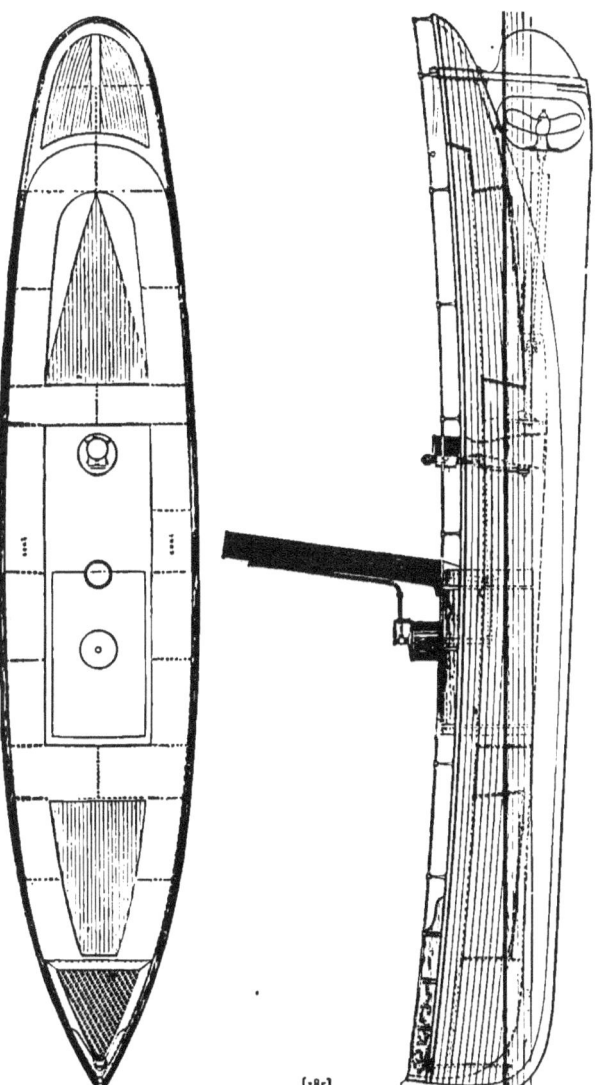

FIG. 69.—STEAM LAUNCH MOHAWK. BUILT BY NEW YORK SAFETY STEAM POWER COMPANY.
28 ft. long; 5 ft. 10 in. beam; 28 in. draft. Power, 3½x15 engine.

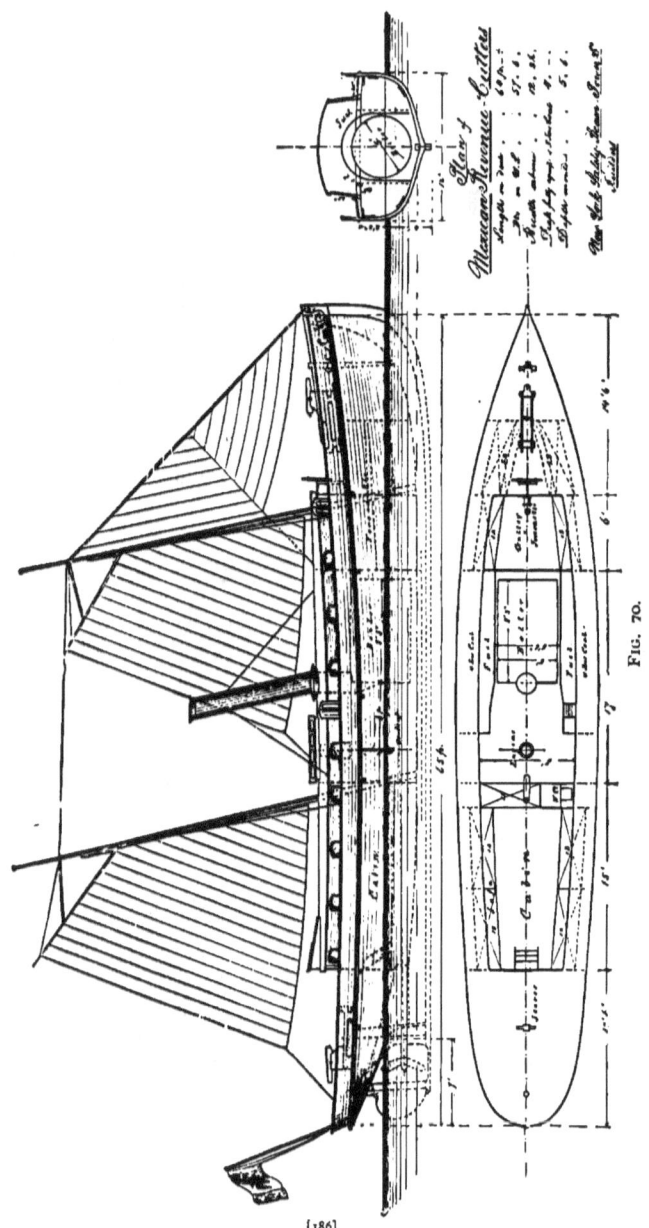

Fig. 70.

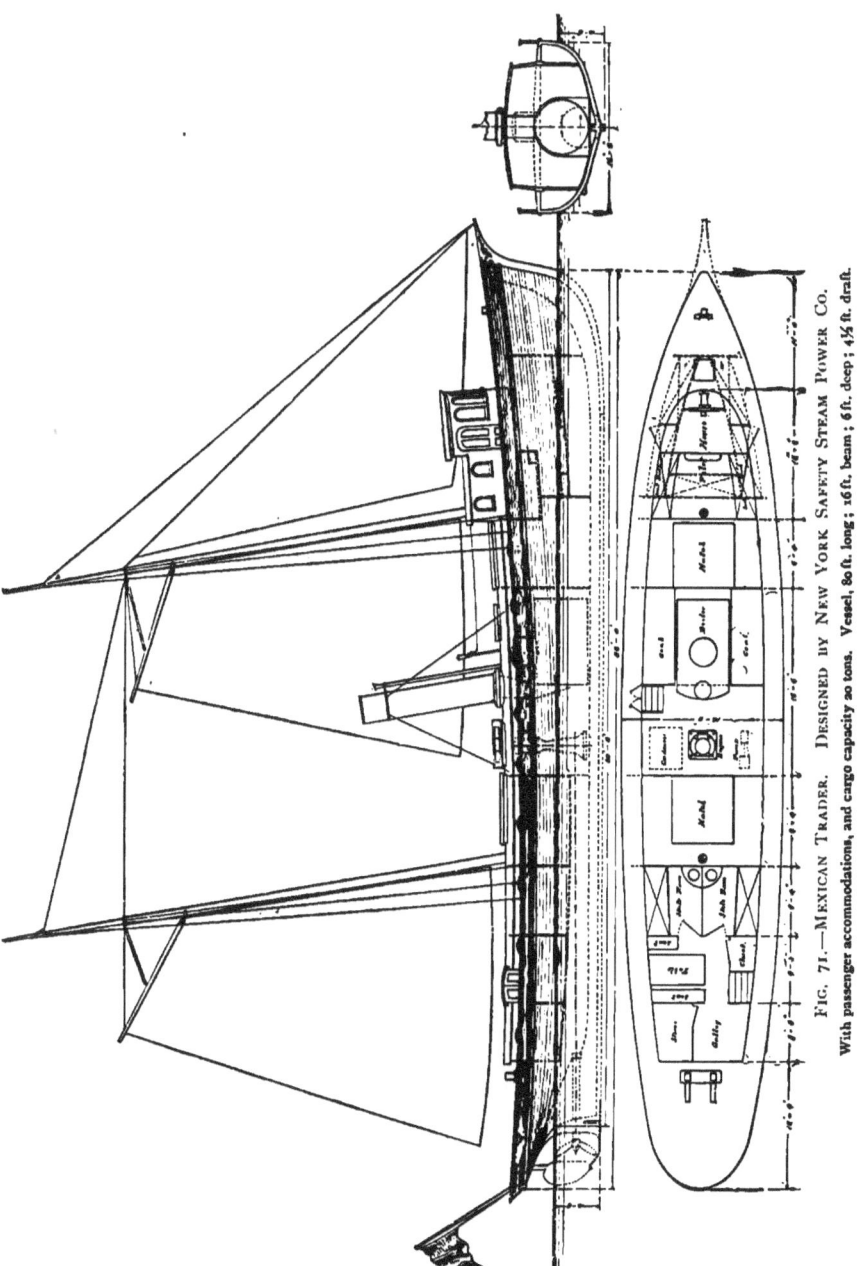

FIG. 71.—MEXICAN TRADER. DESIGNED BY NEW YORK SAFETY STEAM POWER CO. With passenger accommodations, and cargo capacity 20 tons. Vessel, 80 ft. long; 16 ft. beam; 6 ft. deep; 4½ ft. draft.

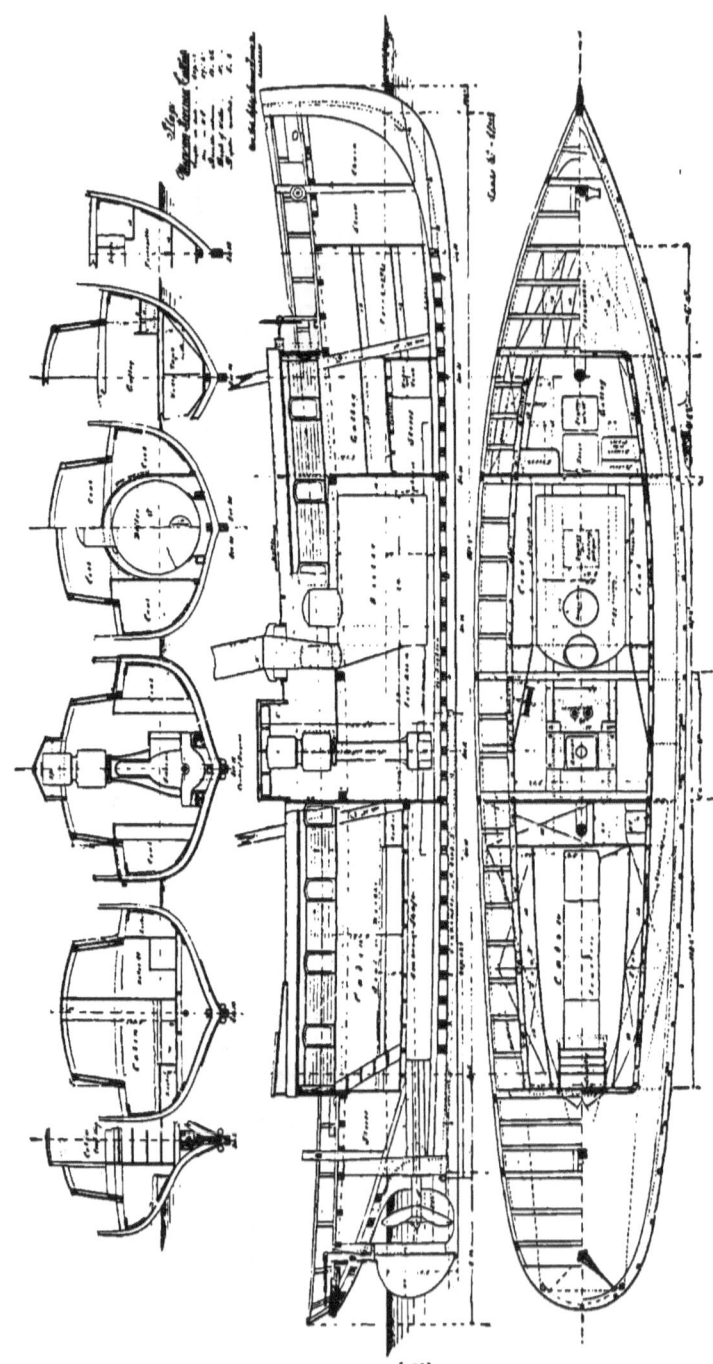

FIG. 72.—GENERAL PLAN OF STEAMERS TAMPICO, CAMPECHE, AND PROGRESO. BUILT BY NEW YORK SAFETY STEAM POWER CO.

Principal Types of Yacht Machinery. 189

The steamers of the Tampico and Mexican Trader class, although not strictly yachts, could be readily altered to such by trifling modifications in their interior arrangements. For cruising the models are quite suitable, being able and roomy and more serviceable than lean craft cut away for the sake of the highest speed. The general plans of these vessels will also be interesting in determining the design of new yachts.

The number of launches and fast yachts on the great fresh-water lakes and Western rivers has increased at a rapid rate during the past years, and a very large fleet, counting into the thousands, is destined to be called into existence with the growing population and wealth of the country. Chicago is the chief center for construction of machinery and hulls of the most approved kind. The Willard high-speed engine has taken a prominent rank, owing to its reliability when working up to 300 turns as a regular working speed. This engine has proportions and material which are the result of the extensive experience of Messrs. C. P. Willard & Co., of Chicago, whose practice in the most popular sizes is shown below:

Diameter Cylinder.	Stroke.	Revolutions.	Boiler Pressure.	Actual Indicated Horse Power.	Height.	Weight.
6 in.	8 in.	350	130 lbs.	40	54 in.	875 lbs.
7 in.	8 in.	350	130 lbs.	60	54 in.	1000 lbs.

The crank shaft is of steel, crank arms being fitted with counter balances bored to fit crank discs. The after journal forms the thrust bearing, the thrust from screw being taken by four steel rings, solid with the shaft. Steam is cut off by a slide valve at half stroke. The illustrations show the engine in place, bolted to the engine keelsons, and also in section. The piston is conical for strength and securing the rod. Cross head is of bronze with the piston rod threaded into it and secured by a jam nut. The boilers built by the same firm have previously been described.

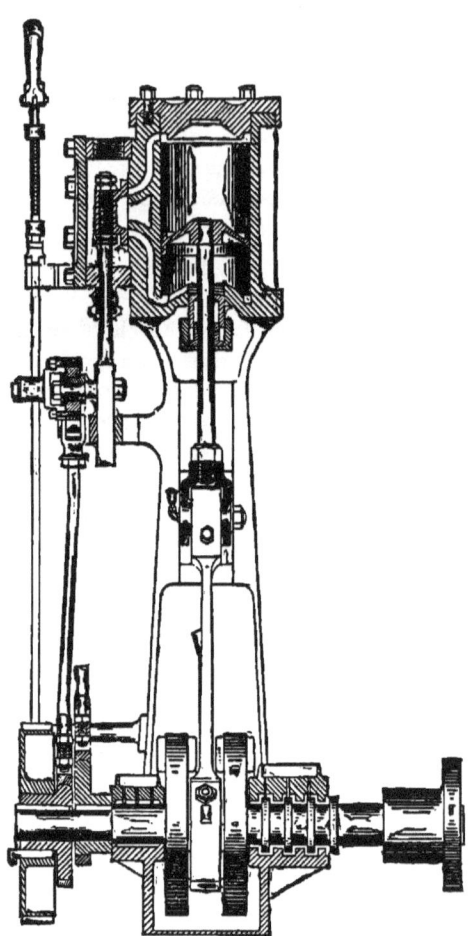

Fig. 73.—The Willard High Speed Marine Engine—Sectional View. Chas. P. Willard & Co., Chicago.

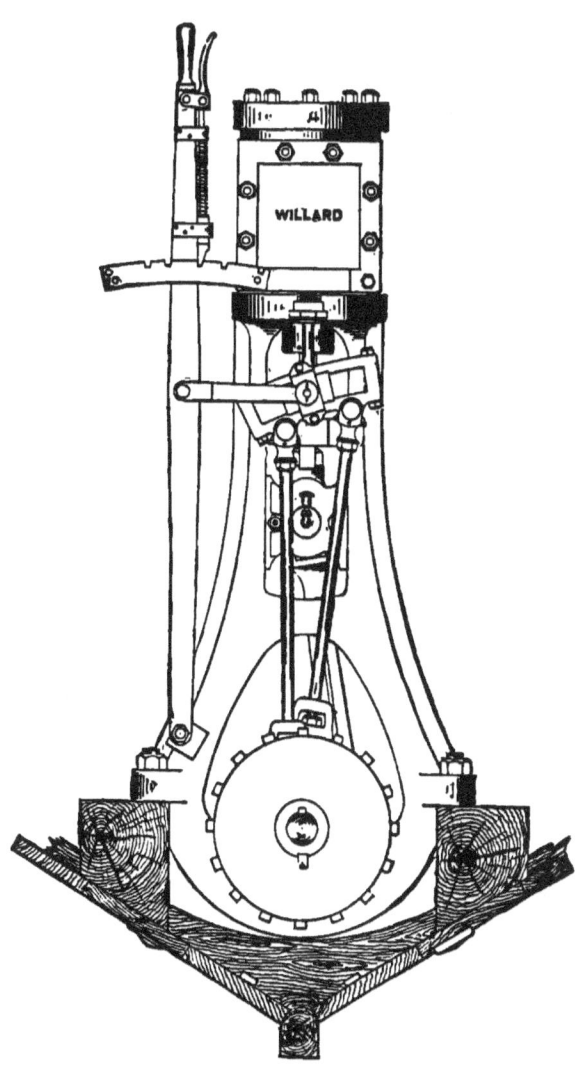

Fig. 74.—The Willard High Speed Marine Engine—Front Elevation, Chas. P. Willard & Co., Chicago.

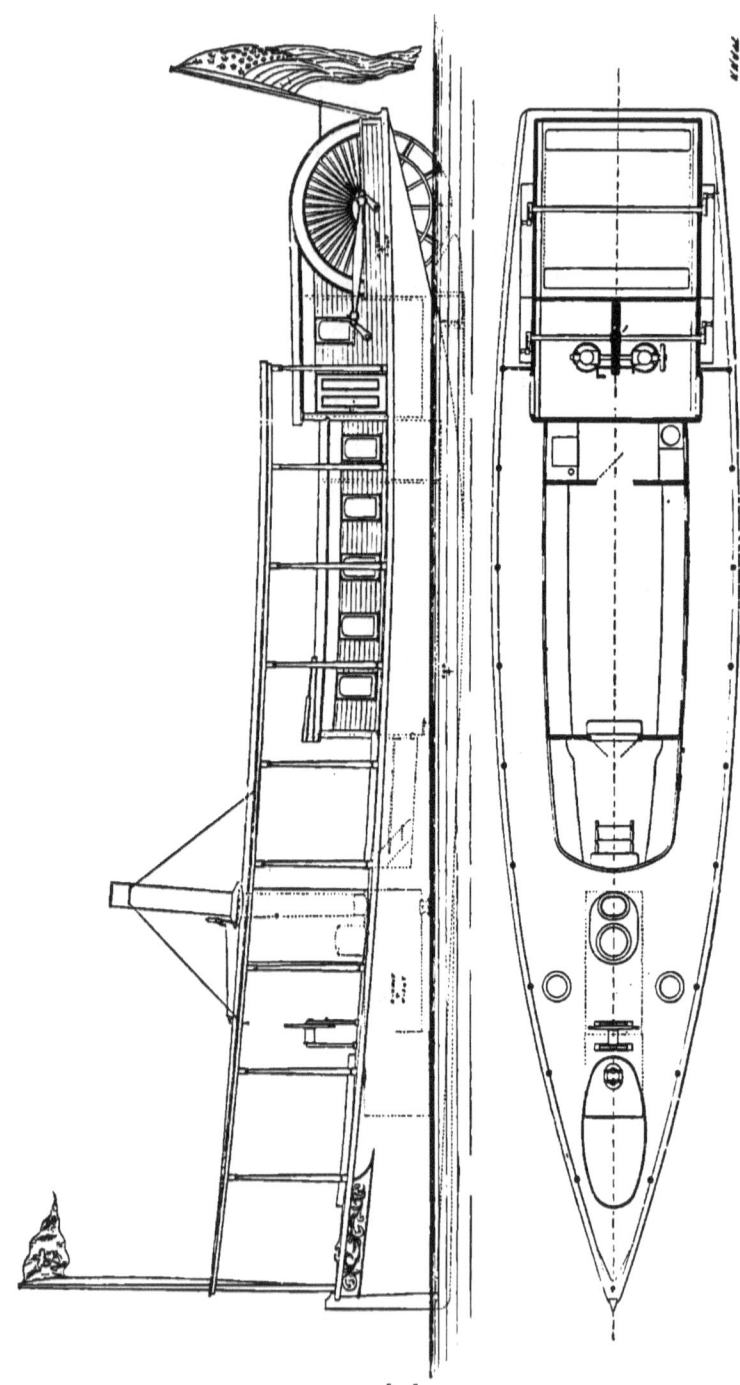

FIG. 75.—STERN-WHEEL STEAM YACHT GRANDE DUCHESSE. BUILT BY NEW YORK SAFETY STEAM POWER CO. 50 ft. long; 12 ft. beam; 4 ft. deep; 20 in. draft. Driven by a pair of Vertical Engines.

Principal Types of Yacht Machinery. 193

The common use of the stern paddle-wheel in the freight and passenger steamers of Western rivers has led to a general appreciation of its merits. It is considered equally as efficient for power or speed as the side-wheels adopted on the steamers of Eastern rivers. For the many thousands of miles of smooth-water navigation throughout the West, the stern-wheel is certain to hold its own in public

FIG. 76.—STERN-WHEEL LAUNCH FOR WESTERN RIVERS.

estimation. In the uncharted headwaters of many of the great river systems, the occurrence of bars, snags and shoal patches renders the stern-wheel better adapted than the screw for light draft in connection with moderate first cost, and it can be applied with less expense, weight or breadth of space than the side-wheel. The general arrangement of launches driven by a wheel over the stern is shown in the accompanying illustration.

The hulls are flat bottom with a round up in the run, and the sides are either straight from bottom to sheer line or have a slight swell amidships and some flare forward according to the moulding of the frame. Such launches can easily be shipped on a flat-car to points of destination. Machinery consists of a submerged-tube vertical boiler, as described in the chapter on boilers and double engines, with horizontal cylinders in the stern, connected direct to the stern-wheel shaft. They are reversed and controlled by a single lever and throttle-valve. The details of a 40 ft. launch built by Willard & Co., are as follows:

Length of hull, feet	33
" boat, " over all	40
Beam, outside, feet	8
Depth, feet	3
Draft, inches	16
Number of persons it will carry	30
Approximate speed, miles per hour	10
" consumption of soft coal, per 10 hours, pounds	400
Actual horse power with 70 lbs. of steam	10
Engines—two in number—diameter of cylinder, inches	5
" " stroke, inches	20
Vertical boiler, diameter, inches	40
" " height, inches	53
Diameter of stern-wheel, feet	7

Sea-going launches do not differ from those intended for river purposes, except in being more or less completely decked for keeping out the sea. A good deal of cruising about the coast and in such open arms of the sea as Chesapeake Bay can be got out of a launch of able proportions.

Cruising launches, not being built solely for speed, are more reasonable in first cost, can be depended upon for strength and a long life, and their engines having a lower velocity of piston than those of racing craft, give much less trouble and annoyance, while the coal account is also less serious an item of expense. One or two hands compose the crew. Fuel can be carried for an extended period, as a minimum displacement is no object; and with ample draft, dead rise and an easy bilge, very fair rough-water qualities may be secured, so that this type is especially adapted for family

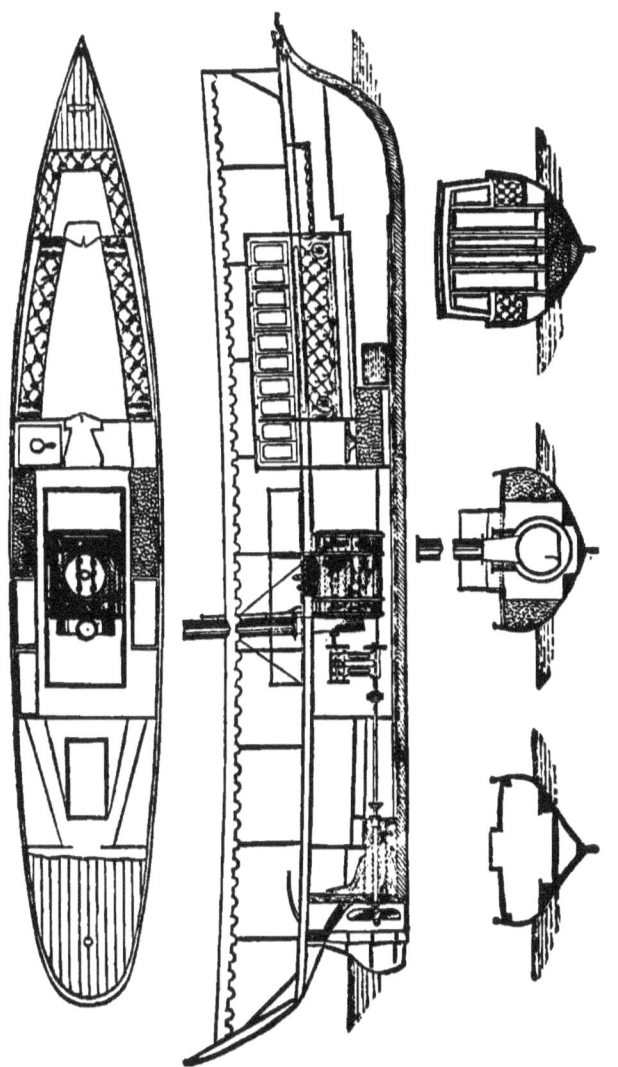

Fig. 77.—Cruising Launch—Fifty Feet Long.

use, for sportsmen, for knocking about, and in short for steam cruising on small tonnage. Persons interested in mechanics, who like to observe the working of machinery, experiment, compare, investigate old theories and perhaps build others in their place, who like besides the romance of adventure, ever-changing scenes and weather, and seek the health of a life in the open air, will do best to invest in a cruising launch and take time and comfort. Racing steamers are good in their way, but they cost more to begin with, and, if you are not in a hurry, offer less in return for your money than the staid but solid cruiser. A fine example is the launch built for an Oriental potentate by J. S. White, of Cowes, England. The boat is decked fore and aft, with the exception of a small cockpit forward. There is a hatch over the boiler space surrounded by a hand-rail and another aft over the quarters for the crew. All these can be battened down with tarpaulins should the craft be caught in bad weather. For use in waters likely to be rough the windows in the cabin can be kept much smaller and every other panel solid, so that if well built, braced and fastened, the danger of being stove in forward is removed. The accommodations, besides cockpit and ample locker room for stores, fishing tackle and shooting outfit, consist of a saloon 12 ft. long, w. c., pantry and galley, situated forward of the machinery. The latter occupies 16 ft., and the remaining length abaft has two berths and storage for the crew. She may be steered by tiller on deck aft or by a wheel on the after cabin bulkhead, a position affording a good view ahead, and protecting the steersman from the weather and sea. This launch made a passage from Cowes to London in the teeth of a strong easterly gale. Her length is 50 ft.; beam, 10 ft.; draft forward, 2 ft. 10 in.; aft, 3 ft. 6 in.; displacement, 11 tons; screw, four blades, diameter 3 ft. 6 in.; pitch, 3 ft. 3 in. to 4 ft. 6 in.; two cylinders, diameter 7¾ in.; stroke, 6 in.; grate surface, 5.5 sq. ft.; heating surface, 215 sq. ft. At the speed of 11.03 miles (9.58 knots) per hour, the number of revolutions was 268, and the gauge showed 76 lbs. per square inch. With a mean effective pressure in the cylinder equal to 75 per cent. of the boiler pressure, the power developed would be 43.4 I. H. P., which must be considered very satisfactory in so small

FIG. 78.—STEAM YACHT FALCON, 166 FT. LONG.

a boat. The hull is built of teak and mahogany, laid in two layers diagonally, coppered and copper fastened.

Along the American coast, where harbors are frequent and easily made, extensive cruises have been undertaken by yachts with high

FIG. 79.—THE FALCON IN THE BAY OF FUNDY.

superstructures. Sufficient buoyancy to hull is a first requisite in this semi-seagoing type, so that the vessel shall lift quickly to a sea, and the deck houses need to be of strong construction, well braced, with wood or iron ports to close the "windows" when the sea is running high. The boiler and engine space should be inclosed with iron or stout wooden bulkheads, so that damage to the superstructure will not flood the stoke hole and put the fires out. The cabins should be built upon a complete deck across the hull, and the hatches for access to the hold must be surrounded by high coamings provided with covers. If the sea can find its way below easily, the risk of swamping would be great. Flush decks are now preferred, as they can easily be had, even in a very small steamer, if she be well proportioned, and at no greater cost than the "cabin houses." For river and harbor yachting, however, the style illustrated by the Falcon has its good points. The accommodations are lofty and cool, and a good view is afforded from the cabin windows or from the elevated position on the upper or hurricane deck. The Falcon, when owned by Mr. R. T. Bush, of Brooklyn, did some extensive coasting in 1884, proceeding from New York to St. John, N. B., and the head of the Bay of Fundy. She was out of sight of land on several occasions, and met rather bad weather, but was reported as going through it in good shape. An idea of her hull can be got from the sketch, showing the steamer left high and dry by the dock of Windsor, the tide falling 60 ft. in the Bay of Fundy, and receding six miles from the dock.

The Falcon is 120 tons burden, 106 ft. long, 16 ft. beam and draws 7 ft. aft.

Regular seagoing cruising yachts are destined to attract more attention in the future than in the past. Hitherto almost every steam yacht in America has been built with the idea of maximum speed uppermost, and seaworthiness has received little consideration except in the largest vessels. In course of time the cruiser will separate from the racer under steam just as he has done in yachts propelled by sail. Only a small fraction of the sailing fleet is given to racing now that such sport involves great expense and efforts strained to the utmost, permitting not a moment's relaxation. Racing under steam is even more extravagant in first cost. A cruising vessel, planned intentionally as such, will boast of less power than the racing machine, hence first cost of engines and boiler will be less. Piston speed will not be so great, and a cheaper grade of machinery will suffice. Bunker capacity must be greater for distant voyages, where fuel is not to be had as readily as between New York and the Maine coast. The space given up to motive power is restricted, and roomy accommodations call for more beam than in the racer. Additional stowage for stores must be supplied in the cruiser, the equipment must be more complete and the rig of serviceable character. The hull must be stronger and show more freeboard. For these reasons the perfect cruiser must be a beamier, deeper and fuller vessel than the racer. She will also be more economical to build, and can be run with fewer hands. Nearly the entire British fleet of steam yachts belongs to the cruiser class, with a speed from 7 to 10 knots for vessels of moderate dimensions, although 11 and 12 is attained in the larger craft. Speed is kept secondary to completeness and adaptability for long voyages. In these cruisers it is possible to obtain very fair internal arrangements and seagoing powers on small dimensions. While the high-speed racer of 60 ft. length is little more than a long launch, lightly decked and half fitted, the cruiser of the same length may be a regular vessel in all that implies.

The Chemcheck, built by Miller, Tupp & Rouse, of London, stands as an example of a large class deservedly popular in England. She was intended for cruising in the Mediterranean. Her extreme length is but 65 ft. with a beam of 12 ft., while the draft of 5 ft. is very moderate. Her construction is such as to fit her for sea work and hard service, the frame being of English oak and American elm well fastened. The planking is of teak coppered, and she is finished throughout—bulwarks, rail, deck houses, skylights and cabins—with the same wood. Her general appearance is that of the ordinary English schooner yacht, the same bow, with round bright bowsprit, high bulwarks (15 in.) and a long counter, while her general proportions, above and below water, are designed to give her sea-going power with good speed under sail alone. The rig, as shown, is that of a two-masted schooner with spars and sails of such size as to be of real service. Of course in so small a boat the bunker space is limited, and coal for long trips cannot be carried, but this disadvantage, attendant on all types of steam yachts of similar size, disappears when the ability to get somewhere under sail is taken into

Fig. 80.—The Cruising Steam Yacht Chemcheyk

account. Her four beams to length give good proportions for sailing, and under the rig she is provided with she is a handy and efficient little vessel under canvas alone. The power provided is a pair of compund engines, surface condensing, with inverted cylinders 9½ and 17 in. by 9½ in. stroke. They are fitted with a valve reversing gear patented by the builders. The air and circulating pumps are driven by an independent engine, a donkey pump is fitted to the boiler, and also a bilge injector. The screw is of 48 in. diameter, 6 ft. pitch, fitted to a 3 in. iron shaft. Steam is provided by a return tubular boiler 60 in. diameter and 7 ft. long, with forty-two 2¼ in. tubes. The boiler is tested to 150 lbs., the working pressure being 60 lbs.

The accommodations are well shown in the plan. Forward is a chain locker, A, and storeroom. The main saloon, B B, is fitted with berths, while by day it serves as cabin and dining room. From it opens a toilet room and w. c., C, and also a pantry, D. Amidships is the engine room F, fitted like the rest in teak, and at the sides the coal bunkers, E E. The ladies' cabin, G, is directly abaft the engine room. It is fitted up with two berths and a skylight. The after portion of the yacht, H H, is fitted up as a galley and quarters for the crew. The elegance and luxury of larger craft are lacking, but the essentials, a staunch little hull, full equipment and cosy quarters, are found in an excellent combination. A still smaller yacht of similar design is the Iris, built in 1883 by the same firm, and also stationed in the Mediterranean. Her length over all is but 60 ft., beam 11 ft., draft 5 ft., depth of hull 6 ft. 6 in. She is built of teak, coppered, and is schooner rigged, being much like the Chemcheck. Her engines are 9 and 16 in. × 9 in., boiler 57 in. diameter and 6 ft. long, screw 48 in. diameter. Her average speed is 10½ knots. She made the voyage through the Bay of Biscay to the Mediterranean and is now stationed at Messina.

A large sea-going yacht of 429 tons, builder's measurement, is shown in the frontispiece.

The Shaugraun was built at Newburg, N. Y., in 1879. She is 169 ft. long over all. 148 ft. 7 in. long on waterline, beam 26 ft. 2 in., depth of hold 12 ft., draft of water 10 ft. 2 in., cubic contents, 26,138.

Engines are double compound surface condensing, with cylinders of 23 and 42 in. diameter and stroke of 30 in. She has two return tubular boilers, each with two furnaces, the total heating surface being 2,000 sq. ft., and working pressure 100 lbs. Propellor is 9 ft. 6 in. diameter. The coal bunkers are of 70 tons capacity. Foremast is 80 ft., foretopmast 44 ft., mainmast 83 ft., topmast 44 ft., bowsprit outboard 32 ft., mainboom 57 ft., foreboom 46 ft. Among her boats she carries a 25 ft. steam pinnace.

DIMENSIONS OF AMERICAN STEAM YACHTS.

Name.	Builder.	Length on deck.	L W L	Beam.	Depth	Draft.	Engines.
Xanthe	Herreshoff, Bristol, R. I.	45	41	9	4.3	2.9	C. I. 2 cyl. 4½ and 7×7. Coil boiler, 3½ ft. diam.
Galatea	P. McGiehan, Pamrapo, N. J.	52	47	10.6	4.4	3.2	C. I. 2 cyl. 6 and 14×9 ft.
Sphinx	John Roach, Chester, Pa.	57	51	12.3	4.3	3	I. 1 cylinder, 11×10. Boiler 5×7½ft.
Amelia	G. Lawley & Son, Boston.	70	65	9	5	3.6	C. I. 2 cyl. 8½ and 15×9. Boiler, 3×8 ft.
Arrow	J. F. Mumm, Brooklyn.	78	76	10	5	5.2	I. 2 cyl. 10×8 ft. Boiler, 5×6 ft.
Stiletto	Herreshoff, Bristol, R. I.	94	90	11.2	8	4.6	C. I. 2 cyl. 12 and 21×12. Pipe boiler, 7×7 ft.
Leila	Herreshoff, Bristol, R. I.	100	95.5	15.4	5.10	6	C. I. 2 cyl. 9 and 16×18. Coil boiler, 7 ft. diam.
Uarda	David Bell, Buffalo.	110	95	17.6	8.6	6.6	C. I. 2 cyl. 14 and 24×16. Boiler, 6×9 ft.
Utowana	John Roach, Chester, Pa.	138	121.6	20.6	11.7	8.2	C. I. 2 cyl. 15 and 28×18. Hor. tub. boiler, 9×11 ft.
Sigma	John Craig, Trenton, Mich.	154	130	21	10	8	C. I. 2 cyl. 16 and 24×28. Boiler, 8×13 ft.
Yosemite	John Roach, Chester. Pa.	182	170	23	18	14	C. I. 2 cyl. 28½ and 40×33. Boilers, 11 and 12 ft.
Stranger	Cramp & Son, Philadelphia.	190	170	23.8	14	10.5	C. I. 2 cyl. 24 and 44×24. Boilers, 10½×11 ft.
Namouna	Ward & Stanton, Newburgh	226.1	217	26.4	15.4	14.3	Tan C. I. 4 cyl. 23 and 42×28. 2 boilers, 11×13 ft.
Nourmahal	Harlan & Hollingsworth, Wilmington, D.	233	221	30	18.7	14.3	C. I. 2 cyl. 34 and 60×36. 4 boilers, 8½ and 10½×12 ft.
Atalanta	Cramp & Son, Philadelphia.	250.3	233.3	26.4	16	13	C. I. 2 cyl. 30 and 60×30. 2 boilers, 10×11 ft.
Alva	Harlan & Hollingsworth, Wilmington, D.	285	252	32	21.6	16.8	C. I. 3 cyl. 32, 45 and 45×42. 2 boilers, 10½×17 ft.

C. I., compound inverted ; Tan. C. I., tandem compound.

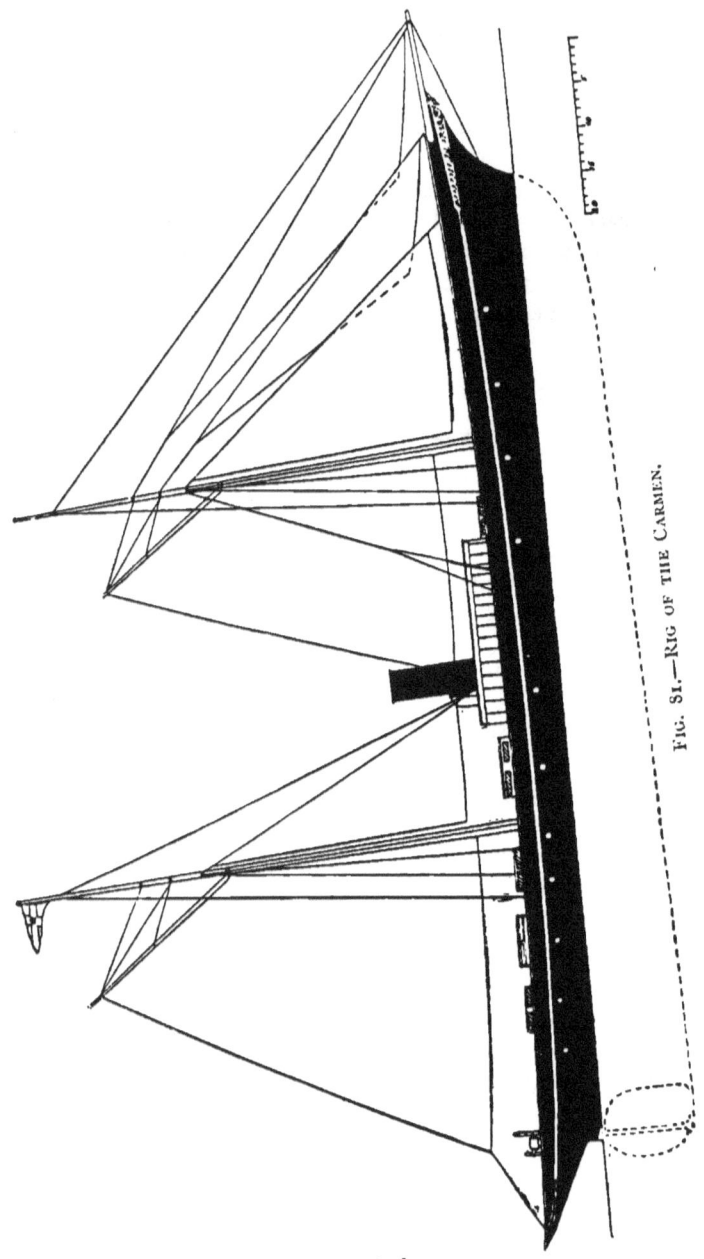

FIG. 81.—RIG OF THE CARMEN.

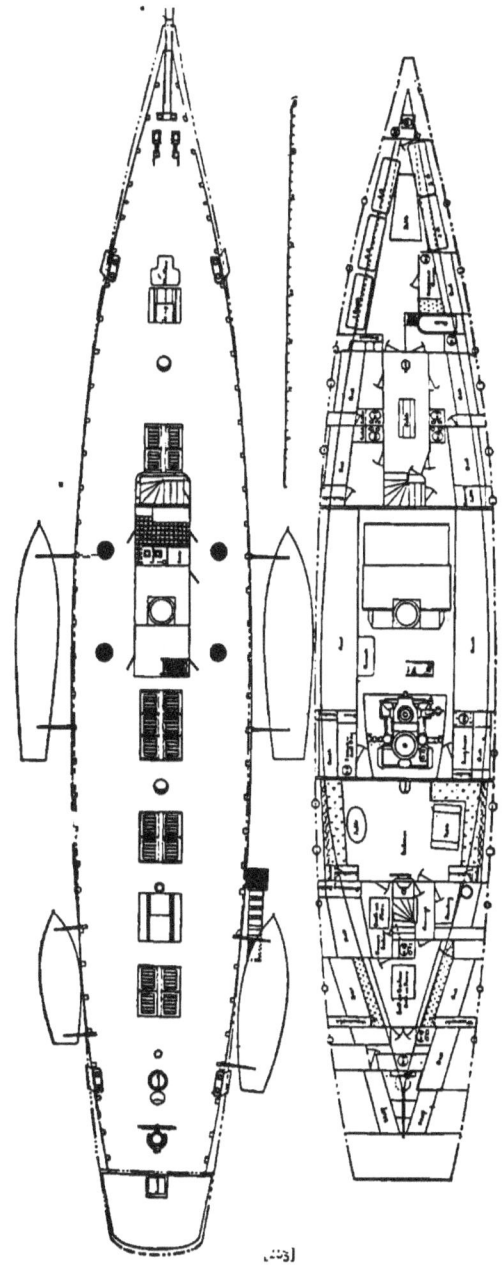

FIG. 82.—CARMEN. DECK AND CABIN PLANS.

A sea-going vessel of English design, capable of long voyages on small fuel consumption and machinery space is the Carmen, designed by Mr. J. Beavor Webb for Sir Thomas Freke, the order being for the most powerful and seaworthy yacht that could be built of 200 tons, a voyage from England to the China Seas being contemplated. Besides some 20 tons of stores, furniture and baggage she was to carry a sufficient amount of coal for a long voyage. Carmen was built of iron by J. Reid & Co., of Glasgow, her engines being built by Walker, Henderson & Co., of the same place. She is 144 ft. over all, 110 ft. waterline, 20 ft. beam and 11 ft. draft. Her displacement is 208 tons, indicated horse power 198, and working pressure 80 lbs. The engines are compounded 14 and 28×31 in. The hull is fitted with three iron bulkheads, one at each end of the engine space and one aft.

The interior arrangements are excellent, both for her owners, guests and crew of fifteen. The latter are berthed in hammock beds in the bow, swung in a large forecastle, forward of which they have a washroom and w. c., while at its after end is the captain's room, neatly fitted, the space under the bed, both here and in all other parts of the ship having large drawers. Abaft the captain's room is a dresser, the galley being above in the deck house. The forward saloon, in which is a dining table, opens into four large staterooms, each fitted with bed, drawers, toilet table, wardrobe and washstand. The boilers and engines occupy no more than their fair share of space, abreast of them being the engineer's and fireman's rooms and bunkers for forty-two tons of coal, sufficient for about 2,700 knots steaming. The main saloon aft of the engines is 11 ft. $\times$ 19, handsomely furnished with a fireplace and mantel, sofas, tables, sideboards and closets, making a pleasant resort in any weather. The owner's cabin is a good-sized room, with bed, toilet table, etc., and with a bathtub below the flooring. The pantry and passage take up the opposite side of the yacht, and further aft is a roomy ladies' cabin, with two berths, two sofas, toilet, etc. Aft of this are closets, store rooms, and a room for the maids. The deck room is large, and affords a fine promenade in good weather. The Carmen carries four boats, a gig 26 ft. $\times$ 4 ft. 3 in., a dinghy 14 ft. $\times$ 4 ft. 6 in., a

Principal Types of Yacht Machinery. 207

cutter 18 ft. 6 in. × 5 ft. 6 in.; and a steam launch 24 ft. × 5 ft. 6 in. The galley and coal box on deck are shown in the upper plan.

She ran from Plymouth to Gibraltar in 4 days 13 hours, thence proceeded to Madeira and Santa Cruz, and from the latter place to Barbados, making the last run in 13 days 4 hours. After a cruise in the West Indies she returned home, running from Bermuda to Holyhead in 16 days. On the trip she proved herself a perfect sea boat. With triple expansion engines her coal consumption would be reduced, and the 42 tons which she carries would serve for about 3,400 knots steaming.

THE WELLS BALANCE ENGINE.

The greater durability, smoothness of operation and lighter construction obtained from balancing reciprocal motion as far as possible is easily understood. The usual method of balancing the weights of the reciprocating parts, by means of counter-weights on the crank, accomplishes only a part of the object desired. It is possible to prevent either horizontal or vertical vibration alone, but impossible to prevent both at the same time. In practice a compromise is therefore adopted with the result that running at a piston speed of 600 ft. a carefully balanced 12 × 12 engine exerts at every stroke an unbalanced force or blow of over a ton. This, of course, would be increased with heavier connections in a larger engine, or a "steeple" compound of the same size.

The relief afforded the crank-pin by an easier cut off, lead or cushioning, does not prevent external vibration, since the elastic steam cushioning transfers the blows received from the piston directly to the cylinder heads. The cumulative effect of these regularly repeated blows is highly injurious and is worthy of serious consideration.

A perfect balance can only be obtained by balancing each motion as is done in the Wells engine, by a corresponding opposite motion of the same weight, exactly in the same plane. In this engine, the reciprocating parts move in opposite directions, and are so proportioned that one piston with all its connections exactly balances in weight the other piston and its connections, producing a perfect equilibrium in the whole circuit of the crank-shaft and at any speed. This permits high speed without vibration.

As the cut of the Wells balance engine will show, the large piston has two rods which pass outside of cylinder A and are connected to the two outside crank-pins, which are set opposite the middle crank-pin connected to the small piston.

Steam enters both cylinders simultaneously, driving the two pistons in opposite directions. The cut represents high-pressure steam, from the steam chest C, forcing down the small piston a, and driving out the exhaust below it into the receiver, while at the same time the low-pressure steam from the receiver is forcing up piston b, and driving the exhaust from above it into the condenser by passage E. In motion, the two pistons approach and recede, the cranks being at 180 deg.

Steam is admitted to both cylinders simultaneously; the pressure against one cylinder head is counteracted by an equal pressure against the other; therefore, no strains are transmitted to the frame or bed of the engine. The steam is so distributed that it exerts an equal force on each piston, but in opposite directions—the thrust of one being perfectly counteracted by the other. The crank power being applied equally at each end of a lever, whose center or fulcrum is the shaft, it will revolve in the main bearings without friction. This will permit the use of high steam pressures without heating.

Steam is distributed by the piston valve c, which is driven by the well known "Joy" valve gear, especially designed to give economical results, by simple and direct connections. The piston valve has been adopted because it is simple, perfectly balanced, and not liable to get out of order. It is more durable than a slide valve under high pressures, and when worn, can be replaced with its casings by duplicates, furnished at a nominal cost, without delay. The valves are proportioned to suit the requirements of the engine, and in some cases have double or treble openings, supplying steam simultaneously.

The engine can be reversed, and the expansion of the steam controlled by set lever e, the same as with a link motion.

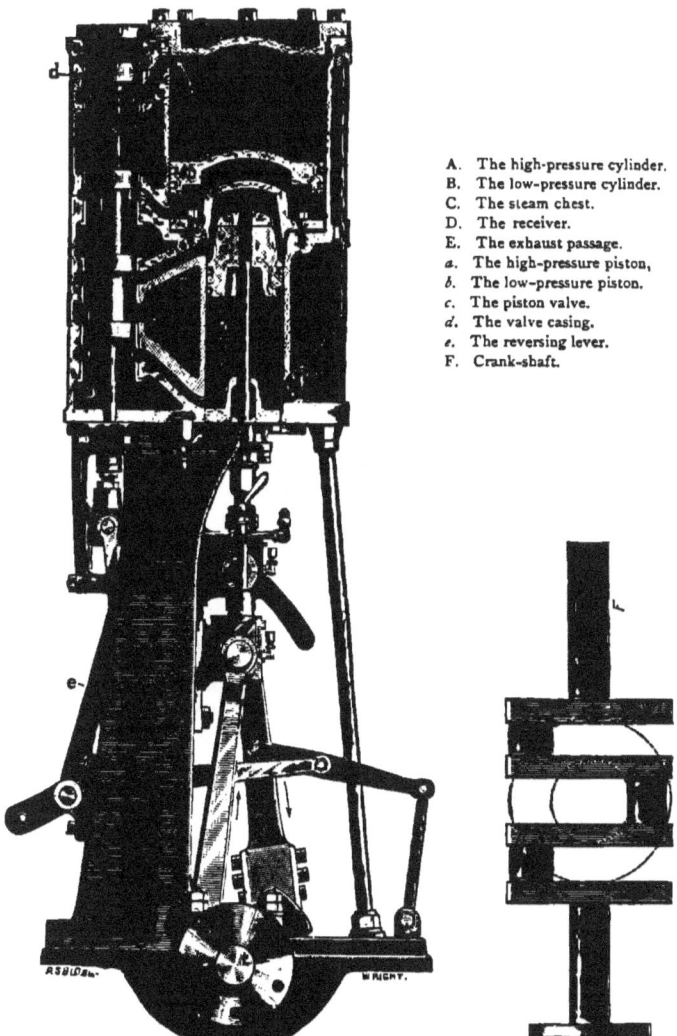

A. The high-pressure cylinder.
B. The low-pressure cylinder.
C. The steam chest.
D. The receiver.
E. The exhaust passage.
a. The high-pressure piston.
b. The low-pressure piston.
c. The piston valve.
d. The valve casing.
e. The reversing lever.
F. Crank-shaft.

FIG. 83.—THE WELLS PATENT COMPOUND BALANCED REVERSING ENGINE.

Balanced weights, balanced momentum, and balanced pressures are qualifications necessary to produce a durable engine. The wear on the shaft journals and their boxes—the most difficult parts to repair—is reduced to a minimum. What little wear takes place on the shaft, due to its weight only, will be evenly distributed around its whole circumference, and not in eccentric form, as is the case in other engines, where it produces additional friction, increased wear, and knocking in the boxes. For these reasons the effect on the main bearing boxes will be highly advantageous.

The crank-shaft is constructed from a block of hammered steel, each crank-pin having a diameter equal to the shaft. Piston rods and valve stem are also of steel and packed with metallic packing. Every part of the engine can be oiled while running.

The low-pressure cylinder receives its steam more direct than in ordinary compounds and hence there is less "drop" in pressure between the two cylinders. High piston speed will also materially reduce cylinder condensation and increase the range of economical expansion. Reduced friction on bearings is a further source of saving, as the following comparison of performance shows, the balanced engine having only half the friction due to pressure, while that due to momentum is entirely removed, so that the bearings will not heat.

	Steeple compound, single crank.	Receiver compound, cranks at right angles.	Balanced compound, cranks set opposite.
Mean pressure on 1st crank-pin	30,000 lbs.	15,000	Two outside pins 7,500 each, or a total of 15,000
Mean pressure on 2d crank-pin	....	15,000	Middle crank pin 15,000
Mean pressure on main bearings	30,000	30,000	......
	60,000	60,000	30,000

Compared to compounds of the ordinary type, it occupies less space, has less weight and fewer parts. The difference in performance of single cylinder condensing engine and compound condensing engine will be made clear from the following figures:

Principal Types of Yacht Machinery.

	Single Cylinder.	Compound.
Size of engine....................	16×16	8 and 16×16
Boiler pressure...................	100 lbs.	100 lbs.
Number of expansions...........	6	6
Force exerted on piston rod at beginning of stroke...........	Single piston 11 tons.	Both pistons combined 5 tons.
Force at end of stroke............	1¼ tons.	1½ tons.
Variation during stroke...........	9¾ tons.	3½ tons.
Variation of temperature due to variation of force or pressure during stroke................	176°	High-pressure cyl. 77° Low-pressure cyl. 99°

That is to say, the single cylinder is subject to twice the variation in temperature and three times the variation in pressure.

Boilers would compare in size and weight and feed water in pounds required per H. P. per hour:

Horse Power.	10 to 35.	40 to 80.	100 up.
Single cylinder exhausting into atmosphere........................	50	40	33
Single cylinder exhausting into condenser........................	50	38	28.
Compound cylinder exhausting into condenser......................	25	20	18

A good boiler will evaporate 30 to 40 lbs. of feed water per H. P. per hour. The Centennial standard is 30 lbs., showing that the above engines, 100 H. P. each, would require boilers as follows:

Single Cylinder Non-condensing	Single Cylinder Condensing.	Compound Condensing.
100 H. P. 33 lbs.	100 28	100 18
300 300	800 200	800 100
30 lbs.) 3300	30) 2800	30) 1800
110 H. P.	93 H. P.	60 H. P.

Showing that the compound requires from one-third to one-half less boiler capacity, which reduces in the same ratio its weight and space, and the fuel consumed per H. P.

The quadruple expansion engine bears the same relation to the compound in the economical use of steam as the latter does to the single-cylinder engine. To obtain greater speed of vessel, more power more economically applied must be employed, without increasing the present weight of machinery or the space it occupies. This can only be attained by carrying higher steam pressures and increasing the number of expansions without undue variation in cylinder temperature. This requires four cylinders. The objection to this type of engine built on the usual plan is the multiplicity of parts, greatly increased friction, weight of engines, and space they occupy. The balance principle eminently fits it for transmitting high pressures that occupy one-half less space, or no more room than the usual fore and aft compound engine. They will also be less in weight, while the weight of boiler will be materially reduced.

To economically transmit 80 H. P. by the Wells system, with a boiler pressure of 175 lbs. and a piston speed of 600 ft., would require a boiler capacity sufficient to supply a cylinder 4 in. diameter and 9 in. stroke, cutting off at half-stroke.

THE COLT DISC ENGINES.

The Colt Disc Engine, made by the Colt Patent Fire Arms Company, is an excellent device which has been supplied to many launches and small yachts. The engine belongs to the self-inclosed class, all the working parts being contained in the cylinder casting, the general shape being such as to suit very readily the form of the boat. It lies so low as to admit, if required, of being floored over so as to economize space, access to it being very seldom necessary, as the lubrication is effected by oil carried in by the steam.

In general construction the Disc Engine consists of six parallel cylinders, cast in a circle like the chamber piece of a revolver. The

FIG. 84.—THE COLT DISC ENGINE.

cylinders (see the illustrations) are open at one end and at the other abut against the steam chest, C, which is separated from the cylinders only by a plate through which the ports are cut. The pistons, A, are in the form of rams or plungers, and when driven home fill up the cylinders.

Facing the open end of the cylinders is a disc, B, which oscillates or rolls on a conical bearing at E, with a ball and socket center. The pistons at this end terminate in blunt, conical points corres-

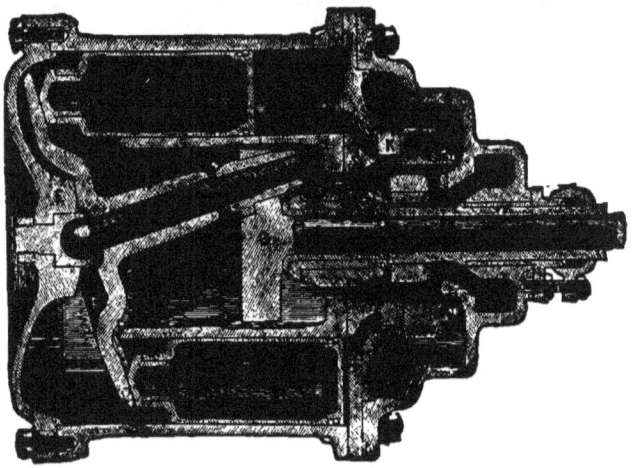

FIG. 85.—SECTION OF CYLINDER CASING.

ponding to the inclination of the disc, which receives its motion from the pistons as they press against it one after another in rotation, the steam being admitted to their opposite ends. The crank G occupies the central space surrounded by the cylinders and is operated by a pin F carried in the center of the disc. The steam distribution is effected by an annular valve, K, surrounding an eccentric, I, inside the steam chest and driven by the shaft H, which passes through it. As the live steam is confined to the space outside the annular valve, nothing but exhaust steam comes in contact with the shaft, so that

Fig. 87.—Interior, Back Cover Off.

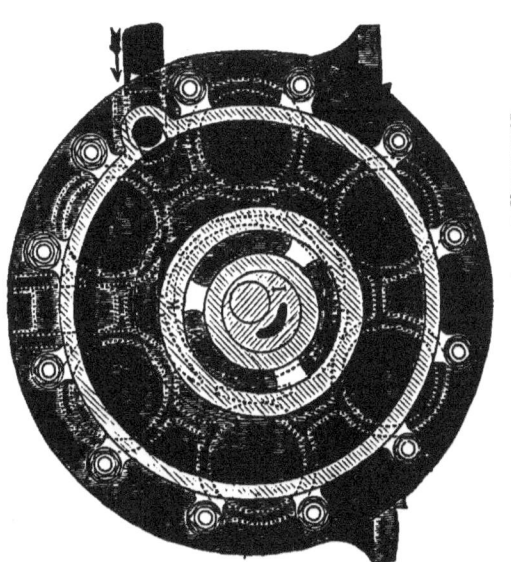

Fig. 86.—Section Across Eccentric.

no stuffing box is necessary where the latter passes through the cover. Of course the pistons are single-acting, the return stroke being effected by the disc forcing them back into the cylinders. As a natural consequence all the strains are continuous, any wear that occurs being followed up by the bearing surfaces, and no pounding is possible. Besides this, in nearly all cases the contact between the working parts is a rolling one, and as there are no mechanical connections anywhere, there is a constant and individual motion of the parts, the result being that while the friction is reduced to a minimum, what little wear there is is distributed uniformly over the entire surface. With the most ordinary care on the part of the engineer one of the engines will run for several seasons without the necessity of spending a dollar for repairs, and when repairs are eventually needed they involve no more than new piston ring or the renewal of some of the bouches, all of which are made interchangeable. The valve arrangement in the disc engine is such as to allow expansion to be taken advantage of to a very great extent with the result of unusual economy in fuel consumption.

A study of the cuts will make the foregoing clear. For that purpose we have produced a section through the engine, in a fore and aft line, another across the eccentric I, showing the ports for steam and exhaust and the manner in which they are opened and closed by the valve ring K, and a third cut giving general interior view with back cover off, and the disc and pistons removed, exhibiting the positions of steam ports and exhaust passages and the crank G. Although usually arranged to cut off at half stroke, a simple alteration in the valve construction enables a much higher degree of expansion to be obtained, but as the peculiar form of the engine renders it unusually suitable for compounding, this method is resorted to where a high degree of expansion is desired. In the perspective view of the engine the thrust collar aft of the crank shaft bearing and the lever for reversing will be noted.

Reversing is accomplished by throwing the bar forward or aft. The bar is forked at its lower end and grasps a collar having a button traveling in a spiral slot in a sleeve connected with the eccentric. By moving the lever, circular motion is imparted to the sleeve

Principal Types of Yacht Machinery.

of the eccentric, which is thrown to the opposite side of the crank, thus causing a reverse motion of the engine.

The following tables are interesting as a guide to selecting the power and corresponding engine required for boats of various sizes, to insure satisfactory performance:

DIMENSIONS, ETC., OF MARINE ENGINES. (STANDARD.)

Effective horse power at 120 lbs. initial pressure.	Total weight.	Diameter of cylinder casting.	Pipe connections.		Proportions of boat and propeller.		
			Steam.	Exhaust	Boat.	Propeller.	
					Length.	Diam.	Pitch.
		In.	In.	In.	Ft	In.	In.
4 to 5	180 lbs.	11	¾	1	25	18 × 18
10 to 12	500 lbs.	15¼	1	1½	35	24 × 24
20 to 25	1,150 lbs.	20⅝	1¼	2	45	32 × 34
35 to 40	2,000 lbs.	25	1¼	2	50	36 × 40
45 to 50	2,800 lbs.	29½	1½	2½	60	42 × 48
65 to 75	3,800 lbs.	32½	1½	2½	75	48 × 54

Propellers up to 32 in. diam. are two-bladed, above that they have three blades.

The Colt boilers are either the return tube or vertical kind, carrying working pressure of 120 lbs.

SPECIFICATIONS OF RETURN TUBE MARINE BOILERS.

	No. 1.	No. 2.	No. 3.	No. 4.
	Ft. In.	Ft. In.	Ft. In.	Ft. In.
Length of boiler........................	5 10	6 4	7	7 6
Diameter of boiler......................	4 2	4 6	5	5 4
Length of furnace and tubes............	4 3	4 9	5 5	5 11
Diameter of furnace....................	2 3	2 6	2 8	3
Number of tubes.......................	55	77	90	75
Diameter of tubes......................	2	2	2	2½
Grate surface, (square feet)............	7.87	10	12	15
Heating surface, effective, (square feet).....	111	162	209	241
Size of disc engine ⎱ Working high pressure	5	6	7	..
for which suitable ⎰ " low "	4	5	6	7

SPECIFICATIONS OF VERTICAL BOILERS.

	No. 1.		No. 2.		No. 3.	
	Ft.	In.	Ft.	In.	Ft.	In.
Diameter of shell.............................	2	10	3	2	3	8
Height of shell................................	4	9	5		5	6
Number of tubes, (all 2 in.)................	55		84		93	
Grate surface, (square feet)................	4.74		6.3		8.29	
Heating surface, (square feet).............	85		139		178	
Size of disc engine ⎱ Working high pressure	3		...		4	
for which suitable ⎰ " low "	2		3		...	

MACHINERY FOR SMALL CRAFT.

Safe machinery, as automatic in action as possible, and of low first cost, has always been in great demand. Recently several new methods and motors have come into prominence, and in many respects supply the necessities of a growing class of boat owners all over the United States.

The Shipman engine and boiler, made by the Shipman Company, of Boston, are well calculated to meet the requirements of launches and steam dinghys, owing to the substitution of petroleum for coal as fuel, the safety of the boiler and the self-acting character of the machinery.

The general operation will be understood from the descriptive illustration of the "Rochester Model" of 1 and 2 H. P. The boiler is of the sectional or pipe variety, each tube being tested to 400 lbs. per square inch, and the completed boiler the same. Although the boiler is practically inexplosive, it is provided with a regular safety valve. A coil pipe heater on top delivers the water to the boiler at 180 deg., the supply being regulated by the float in the float chamber connecting with valve of pump, which opens and closes automatically and keeps a uniform volume of water in the boiler without intervention of the engineer.

Kerosene of 110 deg. test serves for fuel, requiring small storage room, the services of a fireman and the dirt from coal being avoided. The fire is made by the pressure of steam flowing through an atomizer, which throws the kerosene in a very fine spray into the

FIG. 88.—ROCHESTER MODEL.

A. Diaphragm.
B. Pipe connecting diaphragm to atomizer.
C. Atomizer.
D. Oil tank.
E. Lamp or torch.
F. Air pump handle.
G. Pipe connecting air pump to the boiler.
J. Blow-off valve.
K. The drain pipe from exhaust steam heater.
L. The pipe connecting feed water pump to the heater.
M. Drip pipe from the exhaust steam heater.
N. Exhaust steam pipe.
O. Steam gauge.

P. Pop or safety valve.
Q. Water glass.
R. Float chamber.
S. Throttle valve.
T. Swift lubricator to the cylinders.
U. Feed water pump.
V. Strainer to feed water pump.
W. Brass cylinder cap.
X. Shield to the governor.
Y. Steam valve eccentric connected to governor.
Z. The perpendicular rod operated by float in float chamber to cut off the supply of water to the feed water pump.

[219]

fire-box and gives an intense blast without the use of wicks. The combustion is perfect, and the highest results are obtained. The "diaphragm" controls the fire so that a definite pressure can be carried.

The oil tank in front of boiler holds two gallons. This tank has a water space between the oil and fire-box, and this space is filled with water from the feed water supply, and then pumped into the boiler. In this way there is a constantly changing jacket of water three-fourths of an inch thick in front of the oil, making it impossible to heat it.

The feed pump is of brass with lift and force valves. The plunger is connected to the main shaft by an eccentric, and is constantly in operation while the engine is in motion. In connection with the float regulator, it keeps up just the right feed.

The automatic governor increases or decreases the opening of the steam ports, and preserves a uniform speed with varying load. Cylinder has a self oiler, and the connecting rod is also oiled automatically. The piston has adjustable packing rings.

The marine boiler is of wrought iron and with the machinery is operated on the principles explained for the stationary engine. Two eccentrics with link motion and reversing gear are supplied to all yacht engines. About half a gallon of kerosene is required per H. P. per hour, although smaller consumption is reported. This makes the cost about 3½ to 4 cents. The engines have 1, 2, 3 or 4 H. P. A launch of 25 ft. length requires about 2 H. P. for speed of eight miles.

The kerosene, after being thrown into spray by the pressure of the air, before starting is ignited in the fire-box by the "torch." There is little or no smoke. The fire, after the engine is running, is automatically regulated by the "diaphragm." As the pressure of the steam rises to the point at which the diaphragm has been set by a screw, say 100 lbs., the diaphragm is gradually raised, carrying with it a valve which cuts off the passage of steam to the atomizer and thus reduces the fuel supply. If the pressure reaches 100 lbs. the valve entirely cuts off the supply of steam and the flame shrinks to correspond The moment the pressure again falls below 100 lbs.

the valve is released, steam spurts through the atomizers carrying the oil, and the spray is freshly ignited by the torches kept burning at the side of the atomizers.

FIG. 89.—SHIPMAN MARINE ENGINE.

So nice is this action that adding a single lath to the engine's "load" will make an instantaneous and perceptible difference in the fire. If on the other hand the entire load on the engine is suddenly

FIG. 90.—LAUNCH WITH SHIPMAN ENGINE.

removed, and the engine held by a brake, the pressure will not rise more than three pounds, so quickly is the flame choked off.

Under the U. S. laws, quoted in a previous chapter, all boilers which make steam must be inspected and proper papers obtained, and no boiler made wholly or in part of cast iron will be permitted. The Board of Supervising Inspectors of the United States, at the annual meeting in Washington, January, 1887, tested the Shipman engine and boiler and approved the same. A "special license" to run the machinery will be granted to any person familiar with its operation, and the same person can obtain a "special license" as pilot. The address of the local inspector can be obtained at the nearest custom house. Write to him, giving name, length, width and depth of boat and where to be used, also manufacturer's number of engine. The inspector will then test the boiler by hydraulic pressure and issue a permit. If the boat is to be used on inland lakes or rivers not under control of the United States inspection laws, no inspection or license is necessary.

OSCILLATING ENGINES.

The only oscillating engine applied to steam launches in America is the Kriebel valveless engine, built by Rice & Whitacre, of Chicago. This engine has no connecting rod and crosshead, the piston rod actuating the crank without intervening parts, the oscillation of the cylinder providing for the lateral travel of the crank-pin. The motion of the cylinder also serves to open and close the steam and exhaust ports, as will appear from the sectional cut of the engine.

The engine frame, OO, is made in one piece and has boxes on each side to receive the crank shaft, M, and the solid trunnions, E, which project at right angles from the upper head of the cylinder and on which the cylinder is supported and pivoted. The piston, H, is connected by the piston rod, I, to the crank-pin, L, and the three are always in a straight line, consequently as the piston moves up and down, the cylinder vibrates back and forth on the trunnions.

224   *Principal Types of Yacht Machinery.*

The valve, D, is a hollow, cylindrical casting inclosed in a casing, A, attached to the engine frame. The bottom of the valve has a

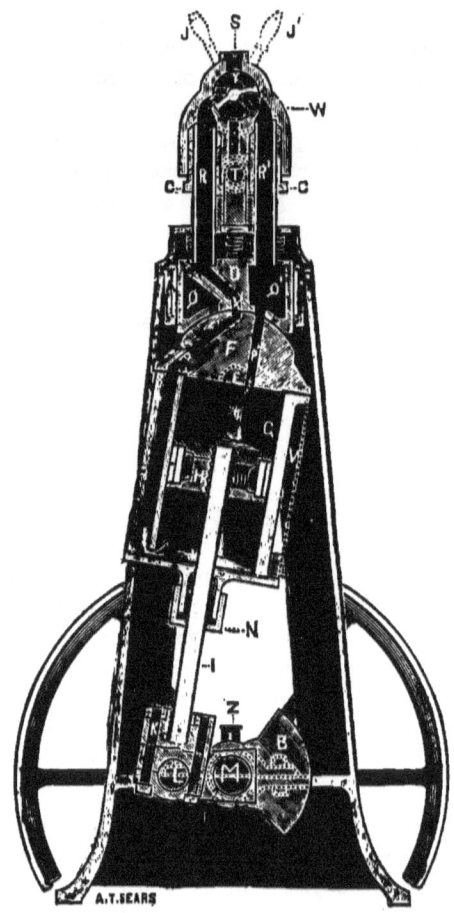

FIG. 91.—SECTION OF KRIEBEL ENGINE.

smooth concave surface, while the upper end of the cylinder, F, has a smooth convex surface. The two surfaces make a perfect joint

Principal Types of Yacht Machinery. 225

and any wear that occurs is automatically taken up by springs, U, coiled around bosses above the valve.

The steam and exhaust pipes, S and T, connect with two brass

FIG. 92.—THE KRIEBEL OSCILLATING LAUNCH ENGINE.

tubes, R and R^1, which are screwed into the valve and communicate with the valve ports, X and QQ^1. There are two cylinder ports, P and P^1, which open into the top and bottom of the cylinder. As the cylinder vibrates back and forth on the trunnions the cylinder ports alternately take steam from the central valve port, X, and exhaust through the ports Q and Q^1. In the reversing engines, the direction of the steam in the tubes can be changed so the cylinder ports will either take steam from the port X and exhaust through the ports Q and Q^1 as above, or else take steam from Q and Q^1 and exhaust through X, and thus reverse the engine.

The piston rod has a long stuffing box, N. The upper ends of the tubes, R and R^1, are received by fixed stuffing boxes, C. B represents a counterbalance, which is bolted to the cranks of engines with 5 × 6 in. cylinder and upward.

The reversing valve on top answers also as a throttle, as by moving the lever to a central position, the steam and exhaust ports are closed and the engine stops. There is of course a saving of friction and weight over engines of the ordinary vertical type having eccentrics, link gear and slide valve.

Piston.		Horse power.	Floor space	Height.		Weight.
Diameter.	Stroke.	With 80 to 110 lbs. boiler pressure.	Occupied by engine bed.	Over all.		Engine and fixtures.
Inches.			Inches.	Feet.	Inches.	Pounds.
3¼ × 4		2 to 3	13 × 12	2	6	160
4 × 4		3 to 4	15 × 15	2	7	200
5 × 6		6 to 8	16 × 18	3	6	400
6 × 7		8 to 10	19 × 20	3	10	560
6½ × 9		10 to 12	22 × 24	4	8	900
7½ × 9		12 to 16	22 × 24	4	8	950

Principal Types of Yacht Machinery. 227

KANE'S PORCUPINE BOILER.

The engine is a simple single cylinder with taper balance plug valve. Five horse power has a diameter of cylinder 5 in., stroke 5 in Kane's Porcupine Boiler is built of lap weld boiler tubes. The five horse power boiler has 118 tubes, 2 in. in diameter by 10 in.

FIG. 93.—KANE'S PORCUPINE BOILER.

long, tapped into a center column which is 7 in. diameter by 38 in. long, like the quills of a porcupine. Within the center column of water is a fire tube which is $3\frac{1}{2}$ in. in diameter and 38 in. long, expanded in the heads of center column. This boiler has 58 sq. ft. of heating surface. Ordinary kerosene oil of from 110 to 150 test is used for fuel. Distillate or residual oil can also be used, and is obtained at any refinery for 3 or 4 cts. per gallon and is found to be very economical. The amount used averages $\frac{1}{2}$ gallon per horse-power per hour, but this can be materially reduced in larger boilers. The oil is stored in a galvanized iron tank, in the bow or stern, and is conveyed from there by $\frac{1}{4}$ in. iron pipe to near the boiler, and thence by a $\frac{1}{16}$ in. pipe to atomizer. The oil is drawn from the tank, atomized by jet of super-heated steam, mixed with sufficient air for combustion and forced into a heated retort. The mixture of oil and steam is decomposed into its componate gases, and conveyed to a burner under the center of boiler where they burn with an intensely hot flame, without smoke or disagreeable noise. No torch is required to keep the fire lit. The burner is not liable to become clogged, and the pitching of the yacht does not affect the fire. The supply of fuel is regulated by an upright lever which controls the flow of oil and regulates the combustion. By turning it to the left more oil can be obtained, and to the right less oil.

NAPHTHA LAUNCHES.

One of the most satisfactory and serviceable of motors recently introduced to take the place of steam in small craft, is the machinery operated by naphtha, as built by the Gas Engine and Power Company, of New York. No steam is used in this motor, no licenses of any sort are required, and explosion is practically impossible. The motor is very compact and takes up little space. The weight is very small, a 2 H. P. engine weighing only 200 lbs; a 4 H. P. 300 lbs., and an 8 H. P. only 600 lbs., which is about one-fifth that of ordinary steam machinery of equal power. Two minutes suffice to get under full headway. Reversing is instantaneous and

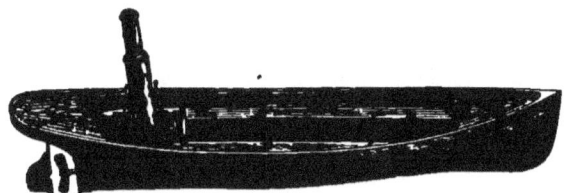

FIG. 94.—NAPHTHA LAUNCH, 18 FT. LONG, 2 H. P.

no attention is needed after the motor is stopped and the boat secured to the dock or hoisted to the davits of a yacht.

The machinery is always clean, as there is neither dirt, ashes nor water. An 18 ft. launch with 2 H. P. engine will carry from six to ten persons, and a 21 ft. boat, with 3 H. P. engine, from ten to fifteen passengers at a speed from 6 to 8 miles an hour, and a cost of six cents per hour. A 30 ft. launch with 4 H. P. will seat twenty-five persons.

The operation of the engine is briefly described as follows: A copper tank in the bow of the boat is filled with 76 deg. deodorized naphtha through a tap screw on top. The tank will hold from 30 to 60 gallons, according to size of boat. A 2 H. P. engine will consume about 3 quarts per hour, and a 4 H. P. engine about 5 quarts.

The vapor which accumulates in the tank is pumped by a few strokes to the burner in the retort or chimney containing a coil of pipe. A flame from a lamp or torch ignites the vapor as it escapes from the burner. This flame supplies the heat for evaporating the naphtha itself, which is next pumped from the bottom of the tank into the coil of the retort. Very little heat is required to volatilize the naphtha in the coil, from which it escapes into the lower casing containing single-acting cylinders operating the crank shaft and

FIG. 95.—LAUNCH SWUNG TO YACHT'S DAVITS.

automatic feed. After expanding in these cylinders the exhaust vapor passes into a condensing pipe and is returned to the tank to be used over again. The flame is supplied as needed by drawing off a small amount of the vapor from the coil, and the only loss of naphtha is represented by the vapor thus burned.

To start the engine: Light alcohol lamp A, and set on rest plate, with tube inserted at bottom of retort; turn air valve B from left to right; give air pump E sufficient number of strokes to force gas

Principal Types of Yacht Machinery. 231

from tank, through outlet pipe to burner, where flame from lamp ignites it and so heats retort. Use air pump one to two minutes in warm weather; but in cold weather much longer, as gas generates very slowly in the tank then. Open wide naphtha valve D, and give five to ten strokes of naphtha pump F, which pumps naphtha from tank in bow to retort on top of engine, and if retort has been sufficiently heated, the pressure will at once be indicated on gauge. Then open injector valve C, which supplies fuel to burner; after

FIG. 96.—30 FT. LAUNCH, WITH FOUR HORSE POWER ENGINE.

which turn reverse wheel G from right to left, or *vice versa*, a few times until engine will run itself. To increase pressure, after engine is started, give a few strokes with naphtha pump F and open wide injector valve C. Regulate both speed and pressure by injector valve C—opening to increase, and closing to reduce speed. To go ahead turn reverse wheel G to left; to back, turn wheel to right. You can reverse instantly, and at full speed. Lamp can be taken out, extinguished, and set on rest plate with tube outside as soon as engine is running. H indicates safety valve. When landing it is only necessary to close injector C and naphtha valve D, fasten your boat, and no further attention is required.

The "air pump" draws the gas from the tank to the burner and

FIG. 97.—NAPHTHA LAUNCH ENGINE.

also to a whistle, by turning valve B from right to left. The naphtha valve should be left wide open to allow free circulation. For the lamp A use alcohol only.

XI.

THE DESIGN OF HULLS.

LITTLE that is precise can be laid down for governing the design of steam yacht hulls. The first requisite is that the displacement at a given draft of water shall be equal to weight of hull, motive power and equipment combined, with ballast added where such is to be carried. In the majority of cases ballast is not necessary to a steam yacht, for the weight of machinery, fuel and stores stand in its stead. But there are occasions where ballast in addition is justifiable. Coal cannot be stowed low, the bunkers reaching up to deck to provide the necessary room, and overhead cylinders will contribute to a high general center of gravity, especially if the rig and deck weights be large also. The resulting top-heaviness can be met in the design by giving the boat more beam, in which case she would be stiff enough without ballast. But the architect may not wish to resort to such correction, for he may prefer a narrower and deeper model to attain other ends in view. Knowing that weight or displacement in itself is not a true measure of resistance, but that larger displacement and cross-sectional area can be driven upon correspondingly smaller beam with like power, owing to the lesser "wave-making" tendencies of narrower hull, the architect may elect to retain small beam and correct want of stability by adding to the displacement a certain amount for an allowance of ballast. The result will be a model of no greater resistance than the wider boat of more beam and less displacement.* At the same time, sufficient stability will be insured by a low center of gravity instead of depending upon the high meta-center due to large beam.

The extra depth, weight and easy form are by some designers preferred for good sea-going qualities and easy behavior. The weight of ballast cannot in such cases be put into greater weight of engines and fuel, because such addition would be in the wrong place for stability and might aggravate the evil of top-heaviness. As a rule, however, steam yachts are planned to do without ballast. The great majority can afford to overlook the highest sea-going qualities, particularly along the American coast, where smooth-water navigation and short runs outside from port to port in reasonably fair weather preponderate greatly.

No directions for proportions of hull can be quoted. In general, five beams to waterline length with such depth as the displacement calls for, will serve the wants of the cruising steamer. For high speed, the ratio between breadth and length is increased. Experience as well as inference teaches that the longest aud narrowest hull is the form of least resistance, and the only restriction is the demand for beam enough to bring about the requisite stability. Thus, the racing shell-boat, propelled by oars, is not built wide and shallow with a saucer section, but on the contrary, the cross section is almost semi-circular and the width of the boat narrowed down to the utmost practicable, the only limit being the width necessary to seat the man pulling the oars. Similar forms would be followed in the hulls of high speed steamers, but for the fact that such forms will capsize, unless sustained by the application of extraneous support, which in the racing shell-boat is derived from the blades of the oars resting on the surface of the water with their handles passing through rowlocks closed across the top, acting as long supporting levers rigged out on each side of the boat. Such assistance being impossible in a steam yacht, more beam is taken in proportion to length, so that the vessel will be able to float on her own bottom.

* See "Small Yachts," pages 46 and 55. (Forest and Stream Publishing Co., New York.)

The Design of Hulls.

There is also one other consideration governing the choice of beam in proportion to length. This is a physical rather than a theoretical restriction. As the length is increased, the "lines" of the hull will of course become finer and more favorable to speed. But the increase in length is also accompanied by an increase of weight of hull and we have to draw more and more upon the displacement to float this weight, which is equivalent to robbing the driving power of an equal amount. While, therefore, form is being refined for speed on the one hand, we are on the other hand diminishing the possibilities for driving power.

Now, up to a certain not well defined point, it is found in practice that more is gained by refining the hull than is lost by the restriction to driving power. Up to that point, it is advantageous to high speed to narrow the hull. But after this point is once passed, a further refinement of hull is no longer beneficial to speed, and the loss in driving power would make itself evident by a loss in speed. The explanation of this limit to narrowing beam is simple enough. When the critical point mentioned has been reached, the lines of the hull will already be extremely sharp. A further diminution of the beam will affect the general angle of entrance and run only very slightly, while the extra length will add very perceptibly to the weight of hull, so that we would be losing in the weight of engine faster than the gain due to the small additional refinement of hull.

Just where the limit to fining of hull really lies, cannot be answered except through experiment. The limit will vary more or less with the form of the hull as a whole and the character of its lines, and to that extent must remain a matter of judgment in each case. Stated broadly, the builders of high speed yachts adhere to seven and eight beams to waterline length, and sometimes even go beyond. These proportions seem fully justified by the well known tendency of beam to throw off waves, representing a loss in power As speed is increased, the five beams of the seven to ten knot cruiser must give way to much narrower bodies in order that they may be driven at fifteen to twenty knot rates.

The depth will be regulated by the beam and contour of midship section, since the required displacement upon fixed length and breadth depends in the main upon the area of the midships.

The high speed torpedo boats of most recent European construction have from eight to nine and a half beams to length, the latter being the extreme beyond which present experience shows no further profit.

The character of cross section varies according to the views and purposes of the designer. Some boats are given a great deal of dead rise to the floor, with flare to the topsides above water. Others are distinguished by flat floor and low bilge, particularly where the draft is to be small. The illustrations throughout these pages supply ample information on this head.

Fore-and-aft waterlines follow no specific rule. Wide boats need some hollow in the ends to produce sharp entrance. Narrow boats are so fine from their dimensions that the entrance is frequently wedge shape or even parabolic in character. Wide boats need greater length of entrance than narrow craft, as the beam has to be "conciliated." In narrow high-speeds the length of run is increased to insure complete closing of the wake and avoid unbalanced "head" at the bow.

Towing competitive models through tanks with adequate instruments for correct notation is the only method upon which the designer can depend for positive forecast of speed performance.

LIGHTS ON STEAM YACHTS.

Reference has been made in the chapter on Sailing Rules to the decision in the case of the Yosemite vs. Vanderbilt, the judge deciding that steam yachts should carry the lights prescribed for inland navigation. The following correspondence in connection with this case will explain itself. Unfortunately it leaves the question in a contradictory state, though sound practical sense is certainly on the side of the Treasury Department decision as given below:

TREASURY DEPARTMENT, WASHINGTON, D. C., April 9, 1887.

Capt. John G. Hulphers, Steamer Wyanoke, Richmond, Va.:

SIR: I am in receipt of your letter of the 8th inst., with newspaper clipping enclosed, containing a synopsis of the recent decision of Judge Andrews, of the Supreme Court of the State of New York, in the Vanderbilt-Yosemite case, wherein it is held, in substance, that an ocean-going steamer should, when navigating inland waters, change the lights required by Rule II., Sec. 4,233 Revised Statutes, to those required upon harbor, lake and inland steamers, by Rule VII., of the same statute, and you ask whether or not, in accordance with Judge Andrews's decision, you are to change the lights of your vessel, an ocean-going steamer, when she is navigating inland waters.

In reply, I have to inform you that this case having been previously presented to me by another correspondent, the subject was referred to the Solicitor of the Treasury for an opinion "whether officers of the Government were hereafter to administer the laws in accordance with Judge Andrews's decision." To which the Solicitor has replied, in letter, dated April 6, addressed to the Secretary of the Treasury, and now on file in this office, "that officers of the Government, in the administration of the navigation laws of the United States, should be governed by the laws of the United States," meaning as understood by this office, that ocean-going steamers, even though such steamers may be incidentally navigated in inland waters or harbors, and notwithstanding the decision referred to, must carry the lights provided in Article 3, "Revised International Rules and Regulations for Preventing Collisions at Sea," approved March 3, 1885.

JAS. A. DUMONT,
Supervising Inspector.

USEFUL INFORMATION.

WATER.

1 gallon U. S. Standard contains 231 cu. in. and weighs 8⅓ lbs.
1 cu. ft. of water measures 7½ gallons and weighs 62.5 lbs.
1 cu. ft. of salt water weighs 64.3 lbs.

Pressure of a column of water at 60° tempr. $\begin{cases} 2.30 \text{ ft. high} - 1 \text{ lb. per sq. in.} \\ 1.129 \text{ ft. high} - 1 \text{ in. of mercury.} \\ 33.86 \text{ ft. high} - \text{atmospheric pressure.} \end{cases}$

A cubic inch of water, evaporated under ordinary atmospheric pressure, is converted into 1,700 cu. in., or in round numbers, 1 cu. ft., and gives a mechanical force equal to raising 2,200 lbs. 1 ft. high.

The height of a column of fresh water equal to a pressure of 1 lb. per sq. in. is usually computed at 2 ft., thus allowing for ordinary friction.

To compute the horse power necessary to raise water to any given height, multiply the total weight of the column of water in pounds by the velocity in feet per minute and divide by 33,000, to which should be added about 25 per cent. for friction, etc.

Water at high temperature cannot be raised to any considerable distance by suction, as the vapor discharged by water so heated follows the receding piston of the pump, and resists the entrance of the water, and is elastic enough to be compressed and distended by the action of the piston without being displaced, thus defeating the service of the pump. Water at boiling temperature must flow to the pump.

Sea water boils at 213.2° $\Big\}$ under atmosphere pressure, and
Fresh water boils at 212° at 124° less in a vacuum.

Every cubic foot of water evaporated in a boiler at the pressure of the atmosphere will heat 2,000 cu. ft. of space to an average temperature of 75°.

1 sq. ft. of steam pipe will warm 200 ft. of space.

FUEL.

1 lb. of coal will evaporate from 7 to 10 lbs. of water.
1 lb. of dry pine wood will evaporate from 4 to 5 lbs. of water.
1 ton of anthracite coal requires a space of 42 cu. ft.
1 ton of bituminous coal requires a space of 44 cu. ft.
150.35 cu. ft. of air are required for the combustion of 1 lb. of coal.

WEIGHT OF VARIOUS SUBSTANCES.

	Cubic foot.	Cubic inch.		Cubic foot.	Cubic inch.
	Lbs.	Lbs.		Lbs.	Lbs.
Cast Iron......	450.55	.2607	White Pine.....	29.56	.0171
Wrought Iron..	489.65	.2816	Yellow Pine....	33.81	.019
Steel..........	489.8	.2834	White Oak.....	45.2	.026
Copper........	555.	.32118	Live Oak......	70.	.040
Lead..........	708.75	.41015	Sand..........	95.	
Brass.........	537.75	.3112	Clay..........	135.	

WEIGHT OF SHEET AND PLATE IRON PER SQUARE FOOT.

No. B. Wire Gauge..........	16	11	7	3	1		
Thickness.................	$\frac{1}{16}$	$\frac{1}{8}$	$\frac{3}{16}$	$\frac{1}{4}$	$\frac{5}{16}$	$\frac{3}{8}$	$\frac{7}{16}$
Pounds....................	2.5	5	7.6	10	12.5	15.2	17.7

TENSILE STRENGTH.

Weight or force necessary to tear asunder 1 sq. in. in pounds bar iron, 60,000; cast iron, 15,000; wrought copper, 34,000; steel, 120,000; copper wire, 61,000; iron wire, 103,000.

NOTE.—The practical value is about ¼ of the above.

CEMENTS FOR STEAM BOILERS, STEAM PIPES, ETC.

Soft Cement.—Red and white lead in oil 4 parts, sifted iron borings 2 to 3 parts.

RUST JOINTS.

For Quick Setting.—1 lb. of salammoniac in powder, 2 lbs. flower of sulphur, 80 lbs. of iron borings; made to paste with water.

Slow Setting.—2 lbs. salammoniac, 1 lb. of sulphur, 20 lbs. of iron borings.

BISHOP'S
Patent Yacht Pump Water Closet,

FOR ABOVE OR BELOW WATER LINE.

No Valves or Cocks to Turn On or Off.

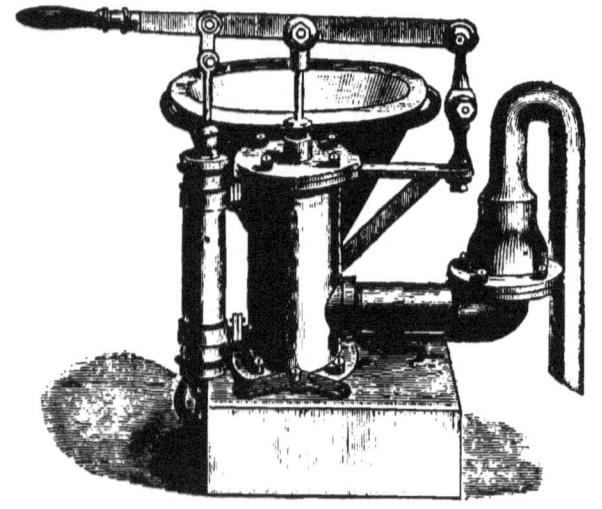

Patented and Manufactured by

WILLIAM BISHOP,

House, Yacht & Ship Plumber,

COPPERSMITH AND STEAM FITTER.

210 SOUTH ST., N. Y

YACHT PLUMBING A SPECIALTY.

CANOE AND YACHT BOOKS.
Published and For Sale by the Forest and Stream Publishing Co.

Canoe and Boat Building.—A complete manual for amateurs. Containing plain and comprehensive directions for the construction of canoes, rowing and sailing boats and hunting craft. By W. P. Stephens, canoeing editor of *Forest and Stream*. With numerous illustrations and twenty-nine plates of working drawings. Cloth, 189 pages, plates in an envelope. Price $1.50.

Canoe Handling.—The Canoe: History, Uses, Limitations and Varieties, Practical Management and Care, and Relative Facts. By C. Bowyer Vaux ("Dot"). Illustrated. Cloth, 168 pages. Price $1.00. A complete manual for the management of the canoe.

Canoe and Camp Cookery.—A Practical Cook Book for Canoeists, Corinthian Sailors and Outers. By "Seneca." Cloth, 96 pages. Price, $1.00. Full and plain instructions about outfit and cooking utensils.

Small Yachts.—Small Yachts: Their Design and Construction, Exemplified by the Ruling Types of Modern Practice. With numerous plates and illustrations. By C. P. Kunhardt. Cloth, 370 pages of type and illustrations and 70 plates. Size of page 14½x12¾. Price, $7.00. This book is intended to cover the field of small yachts, with special regard to their design, construction, equipment and keep. Among the plates will be found many famous and well-known vessels, illustrated with great detail and finish.

Steam Yachts and Launches.—Their Machinery and Management. By C. P. Kunhardt. With plates and many illustrations. Cloth, 250 pages. Price, $3.00.

Yachts, Boats and Canoes.—With special chapters on model yachts and singlehanded sailing. By C. Stansfield-Hicks. Numerous illustrations and diagrams, and working drawings of model yachts and various small craft suitable for amateurs. Cloth. Price, $3.50.

Knots, Ties and Splices.—A handbook for seafarers, travelers and all who use cordage. By J. Tom Burgess. Illustrated. Cloth, 101 pages. Price, 50 cents. Gives all the useful knots and illustrates them intelligently.

The Canoe Aurora.—A Cruise from the Adirondacks to the Gulf. By Dr. Chas. A. Neide, ex-Secretary of the American Canoe Association. Cloth, 215 pages, with map of the route. Price, $1.

Four Months in a Sneakbox.—A boat voyage of 2,600 miles down the Ohio and Mississippi Rivers, and along the Gulf of Mexico. By Nathaniel H. Bishop. With diagram of sneakbox and other illustrations. Cloth, 322 pages. Price, $1.50.

Woods and Lakes of Maine.—A trip from Moosehead Lake to New Brunswick, with large map of Moosehead Lake and Northern Maine, with soundings in Moosehead Lake. By Lucius L. Hubbard. Handsomely illustrated and bound, 223 pages. Price, $3.

Canvas Canoes; How to Build Them.—A complete manual of instruction for building cheap, safe canvas canoes. By Parker B. Field. With a plan and all dimensions, and other illustrations. Paper, 48 pages. Price, 50 cents.

Model Yachts and Boats.—Their Designing, Making and Sailing. By J. du V. Grosvenor. Illustrated with 121 designs and working drawings. Leatherette, 183 pages. Price, $2.00.

Woodcraft.—By "Nessmuk." Cloth, 160 pages. Illustrated. Price, $1.00. A book written for the instruction and guidance of those who go for pleasure to the woods.

OUTDOOR BOOKS.—The Forest and Stream Publishing Co.'s descriptive Catalogue of books on Shooting, Angling, Camping and Outdoor Life, will be sent free to any address.

Forest and Stream Publishing Co., 40 Park Row, New York.

HOUSTON & WOODBRIDGE,

ENGINEERS

AND

Iron Ship Builders,

STEAM AND SAILING YACHTS.

P. O. Linwood, Delaware Co., Pa.

J. BEAVOR-WEBB,
45 BROADWAY, NEW YORK.

AGENT FOR
FOX'S PATENT
CORRUGATED
Boiler Furnaces
Manufactured by
The Leeds Forge Co.
OF ENGLAND.

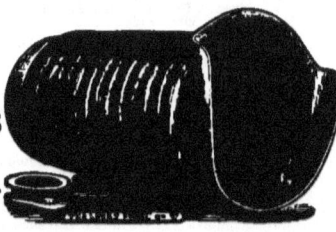

**Flanged Plates,
Boiler Plates**
AND
Blooms.
Best Yorkshire Steel
(SIEMENS').

FITTED IN THE AMERICAN STEAM YACHTS
"Corsair" "Stranger" "Peerless" "Susquehanna" "Alva."
AND OTHERS.

ALSO AGENT FOR
Muir & Caldwell's Steam Steering Gear.
VICKER'S STEEL SHAFTING.

Steam Yachts and Pleasure Boats.
PETROLEUM AS FUEL.

Seven sizes of Steam Yachts, 19 to 40 ft. long. Designed for Yacht Tenders, Pleasure, Cruising and Hunting purposes. Fine models and workmanship. Light, speedy and seaworthy. No noise, smoke or dirt. No danger. No smell. Perfect combustion. Petroleum as fuel. Economical and clean. Easy to operate. Marine and Stationary Engines and Boilers. Yacht fittings. *SEND STAMPS FOR CATALOGUE.*

Celebrated Racine Boats and Canoes. Veneer Canoes with life compartments. Hunting and Fishing Boats. Cedar Lapstreak Boats. Boat Fittings.

SEND STAMPS FOR CATALOGUE.

THOMAS KANE & CO.,
137 & 139 Wabash Avenue,   Racine, Wisconsin,
CHICAGO, ILL.   U. S. A.

Four Unrivaled Products for Yachtsmen.

DIXON'S
"Potlead" for Yacht Bottoms.

Specially prepared for the purpose, and unequalled for Purity and Uniformity of Grain.

"Pot leading" is of value in proportion to the quality of the article used. DIXON'S BLACK LEAD is a pure graphite ground to a fine and even grain, so that there is no waste; and the vessel's bottom treated with it will be of surprising smoothness. It will also be found a protection to the bottom.

DIXON'S SILICA GRAPHITE PAINT.

It is a mixture of perfected graphite and pure linseed oil, so thoroughly mixed that, when applied with the brush, the iron receives a coating of the thin flakes of graphite, and this coating once secured, the metal will stand any heat it will ever receive. For smoke-stacks, boiler fronts, wire cables and all metal work it is without an equal.

Dixon's Graphite Grease.

For gears, for loose-fitting journals and bearings, or indeed for any friction surfaces whatever where the conditions are such that a grease can be introduced, we guarantee perfect usefulness. A little of this grease does a great deal of work.

Dixon's Dry Graphite.

Dixon's water-dressed Dry Foliated American Graphite is a little thin flake of graphite of extraordinary properties. It has unrivaled smoothness and endurance. Its superiority as a lubricant has been attested by all recent writers on friction.

Its enduring qualities are several times greater than those of any oil. Unlike either oil or grease, it is not affected by heat, cold, steam, acids, etc., and acts equally well under the most varying conditions of temperature and moisture.

JOS. DIXON CRUCIBLE CO.,
JERSEY CITY, N. J.

New York Office, 68 Reade Street.

COLT DISC MARINE ENGINE.
NOISELESS and ECONOMICAL.

BEST PROPELLER ENGINE IN THE WORLD FOR
YACHTS, TUGS AND LAUNCHES.

It occupies small space, runs at high speed, has no dead centers, is self-inclosed, has few wearing parts, uniform wear. Constructed in the best manner and of the best materials. Easily operated by any one.

Dimensions, Etc., of Marine Engines (Standard).

Diameter of Pistons	Total Weight	Diameter of Cylinder	Casting over all	Pipe Connections		Proportions of Boat and Propeller for which Engine is Suited.						Prices include Reversing Gear, Thrust Block, Coupling and Automatic Lubricator.
				Steam.	Exhaust.	BOAT.			PROPELLER.			
						Length.	Beam.	Draft.	Diam.	Pitch.		
In.	Lbs.	In.	In.	In.	In.	Ft.	Ft.	In.	In.			
2	180	11	¾	1	25	6	25	18×18		Two Bladed.	$235.00
3	500	15¼	1	1½	35	8	30	24×24			327.00
4	1150	20⅝	1¼	2	45	9	35	32×34		Three Bladed.	475.00
5	2000	25	1¼	2	50	10	40	36×40			735.00
6	2800	29½	1½	2½	60	12	50	42×48			1,055.00
7	3800	32½	1½	2½	75	14	60	48×54			1,370.00

Colt's Patent Fire Arms M'f'g Co.,
HARTFORD, CONN.

ST. MARY'S, OHIO. } OAR MILLS. NEPTUNE ANCHOR WORKS.
MONTPELIER, OHIO. }

DE GRAUW, AYMAR & CO.,

Manufacturers and Importers of

CORDAGE, OAKUM, WIRE ROPE,

Chains, Anchors, Oars, Blocks,

Buntings, Flags,

Cotton and Flax Ducks,

Russia Bolt Rope.

Marine Hardware and Ship Chandlers' Goods Generally.

Nos. 34 AND 35 SOUTH STREET,
NEW YORK.

Small Steam Yachts
AND
STEAM LAUNCHES.
OF EITHER
WOOD OR STEEL.

MACHINERY.

High Pressure Non-Condensing, Compound Non-Condensing, Compound Jet and Surface Condensing, Triple Expansion Surface Condensing.

Our boats are not experimental, but are powerful, fast, and economical of fuel. Burn either coal or wood. Can refer to them in successful operation in all parts of the United States. Illustrated catalogue, including engines, boilers, propeller wheels, also six photographs of finished launches, sent on receipt of 12c. in stamps.

CHAS. P. WILLARD & CO.,
236 Randolph Street, Chicago, Ill.

www.ingramcontent.com/pod-product-compliance
Lightning Source LLC
Chambersburg PA
CBHW020806230426
43666CB00007B/886